My Book

This book belongs to

Cover Design by :
Gowri Vemuri

First Edition :
December , 2019

Second Edition:

Apriil,2020

Author :
Gowri Vemuri

Edited by :
Raksha Pothapragada
Ritvik Pothapragada
Questions: mathknots.help@gmail.com

NOTE : VDOE is neither affiliated nor sponsors or endorses this product.

Dedication

This book is dedicated to:

My Mom, who is my best critic, guide and supporter.

To what I am today, and what I am going to become tomorrow,

is all because of your blessings, unconditional affection and support.

This book is dedicated to the

strongest women of my life ,

my dearest mom

and

to all those moms in this universe.

G.V.

Visit www.a4ace.com

Also available more time based practice tests on subscription
Subscribe to Math-Knots you tube channel for concept videos
E-mail us: mathknots.help@gmail for more concept based videos with a proof of purchase.

The Virginia Board of Education and Virginia Department of Education (VDOE) have developed the Virginia Assessment Program (VAP) to measure and evaluate students' academic progress in the Standards of Learning (SOLs). The SOLs indicate Virginia's expectations for what students should know and be able to do in the subject areas of reading, writing, mathematics, science, and history/social science.

Students in grades 3-12 to take the Standards of Learning (SOL) assessments each year. Some of the tests are required by all students each year, and others are required only at specific grade levels. Additionally, with the removal of some SOL tests in recent years, the VDOE assigned the responsibility of the creation and administration of alternate, performance - based assessments on local divisions. Student scores from these tests determine a school's and the division's state accreditation and measures progress toward meeting federal targets.

Virginia Standards of Learning (SOL) tests are generally given online unless a student has an identified and documented need to be assessed using paper,pencil format. The test question format is typically multiple choice, and each test contains some technology enhanced items.

GRADE 3-8 STANDARDS OF LEARNING(SOL) TESTS

GRADE 3	GRADE 4	GRADE 5	GRADE 6	GRADE 7	GRADE 8
	VIRGINIA STUDIES				WRITING
MATH	MATH	MATH	MATH	MATH	MATH
READING	READING	READING	READING	READING	READING
		SCIENCE			SCIENCE

NOTE : VDOE is neither affiliated nor sponsors or endorses this product.

END OF COURSE STANDARDS OF LEARNING (SOL) TESTS

GRADE 9	GRADE 10	GRADE 11
ALGEBRA I	GEOMETRY	ALGEBRA II
EARTH SCIENCE	BIOLOGY	CHEMISTRY
WORLD HISTORY I	WORLD HISTORY II	VIRGINIA & US HSTORY
		WORLD GEOGRAPHY
		ENGLISH : READING
		ENGLISH : WRITING

Any Student taking one of the courses listed here is expected to take the corresponding end-of-course SOL test. The grade levels depicted here represent grade level at which students typically participate in these courses.

NOTE : VDOE is neither affiliated nor sponsors or endorses this product.

SOL Test Scoring and Performance Reports:

Standards of Learning assessments in English reading, mathematics, science and history/social science are made up of 35-50 items or questions that measure content knowledge, scientific and mathematical processes, reasoning and critical thinking skills. English writing skills are measured with a two-part assessment that includes multiple-choice items and an essay.

Student performance is graded on a scale of 0-600 with 400 representing the minimum level of acceptable proficiency and 500 representing advanced proficiency. On English reading and mathematics tests, the Board of Education has defined three levels of student achievement: basic, proficient, and advanced, with basic describing progress towards proficiency.

Performance Achievement Levels:

- The achievement levels for grades 3-8 reading and mathematics tests are: *Pass/Advanced*, *Pass/Proficient*, *Fail/Basic*, and *Fail/Below Basic*.
- The achievement levels for science tests, history tests, and End-of-Course (EOC) tests* are: *Pass/Advanced*, *Pass/Proficient*, and *Fail/Does Not Meet*.
- The EOC Writing (2010 SOL) test, EOC Reading (2010 SOL) test, and EOC Algebra II (2009 SOL) test have an achievement level of *Advanced/College Path* in place of the *Pass/Advanced* achievement level.

NOTE : VDOE is neither affiliated nor sponsors or endorses this product.

INDEX

FORMULA SHEET

1. Area of a triangle

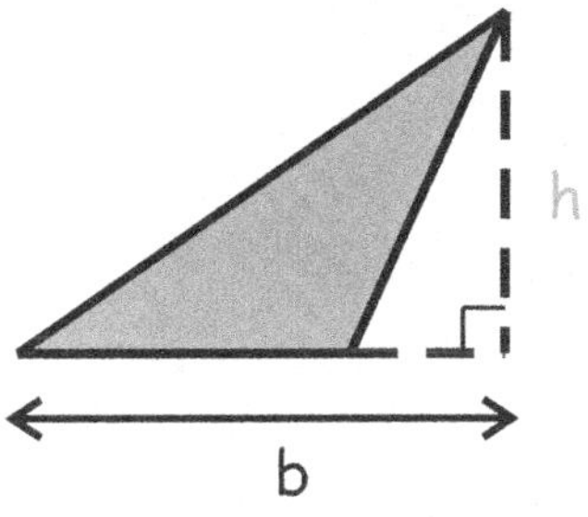

$A = \frac{1}{2} bh$

2. Area of a parellelogram

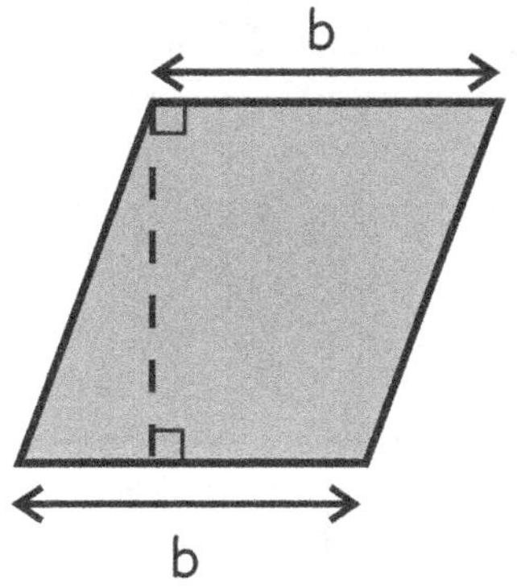

$A = bh$

3. Volume and Surface area of a cuboid

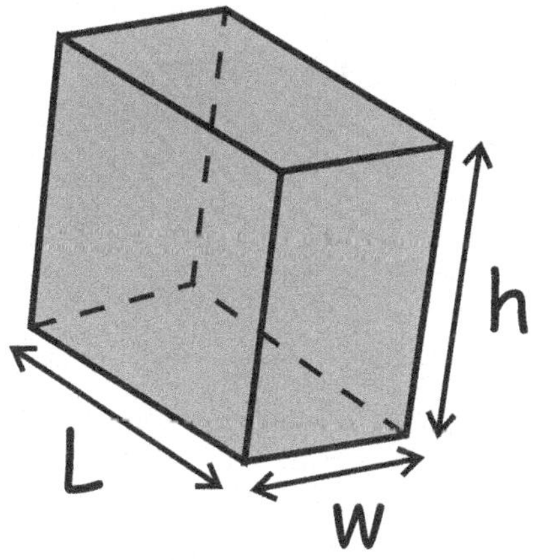

$V = lwh$

$S.A = 2(lw + lh + wh)$

4. Perimeter and Area of a Square

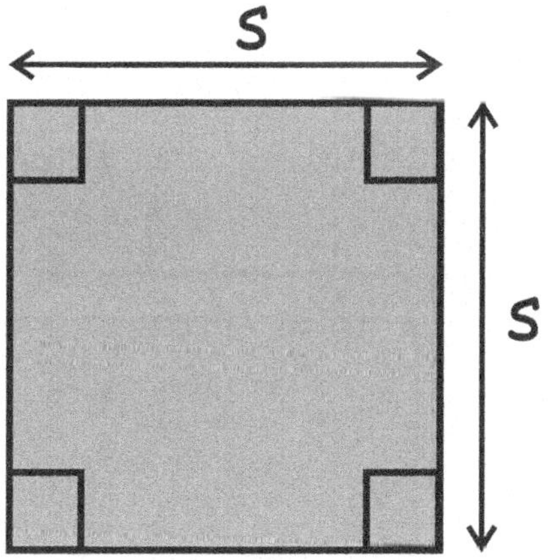

$P = 4s$

$A = s^2$

5. Area of a Trapezium

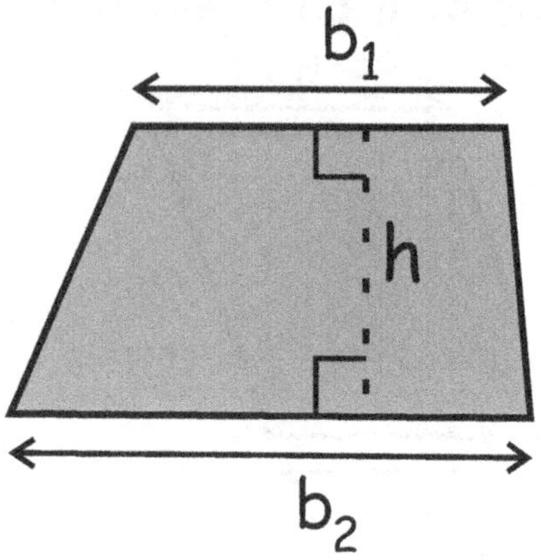

$$A = \frac{1}{2} h(b_1 + b_2)$$

6. Circumference and Area of a Circle

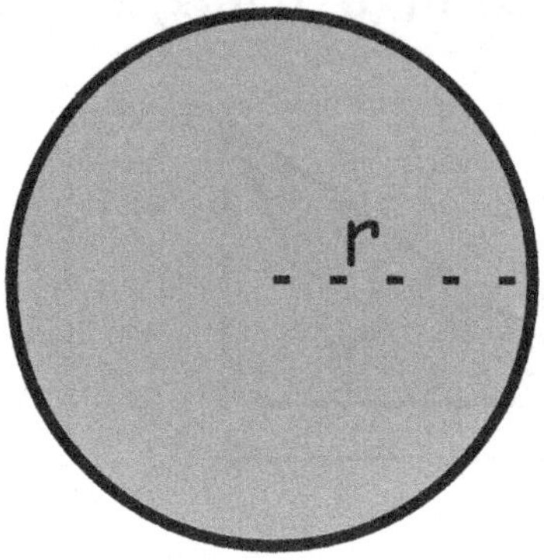

$$c = 2\pi r$$
$$A = \pi r^2$$

pi

$$\pi = 3.14$$
$$\pi = \frac{22}{7}$$

7. Perimeter and Area of a Rectangle

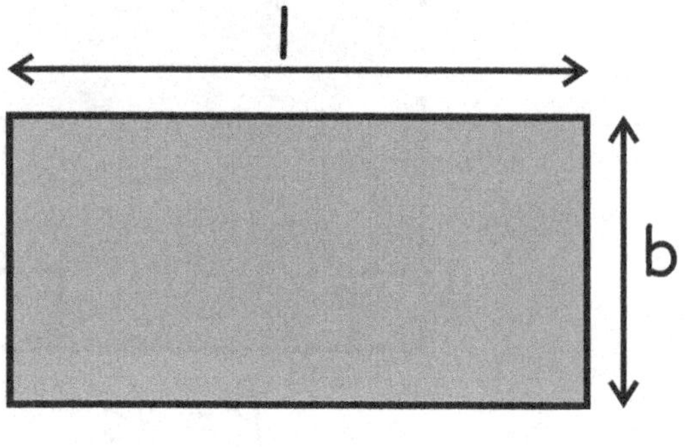

Area = l b ×

Perimeter = 2(l + b)

8. Acute angle

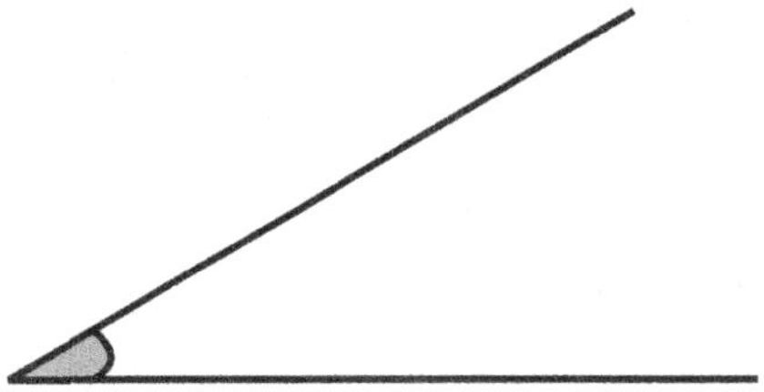

Any angle less than 90^0
is called as acute angled

9. Right angle

An angle equal to 90^0
is called as right angled

10. Obtuse angle

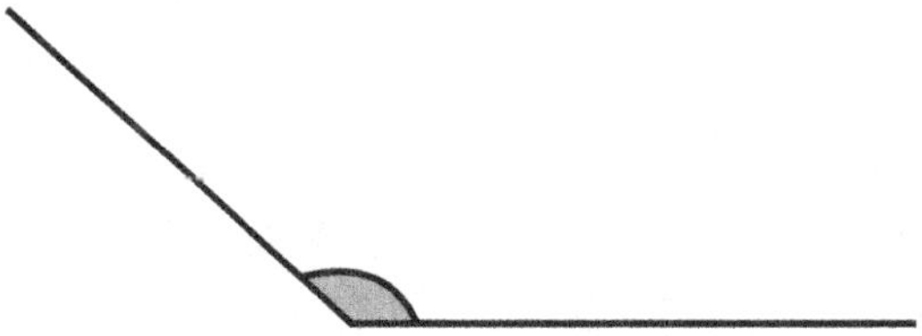

Any angle greater than 90^0
is called as obtuse angled

11. Perimeter of a polygon KLMNOP

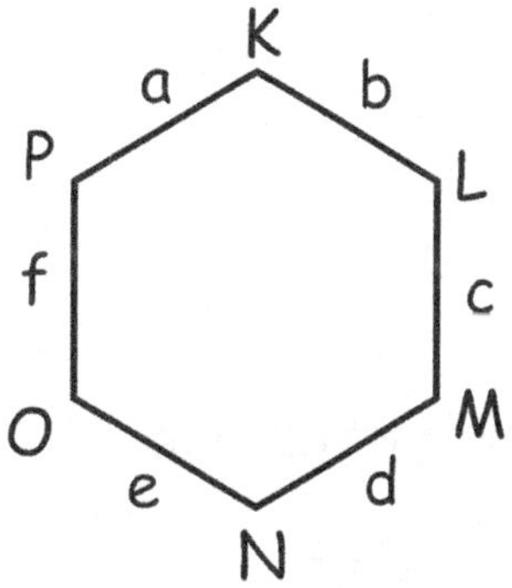

Perimeter =
(a + b + c + d + e + f) units

Abbreviations

milligram	mg
gram	g
kilogram	kg
milliliter	mL
liter	L
kiloliter	kL
millimeter	mm
centimeter	cm
meter	m
kilometer	km
square centimeter	cm^2
cubic centimeter	cm^3

year	yr
month	mon
hour	hr
minute	min
second	sec

volume	V
total Square Area	S.A
area of base	B
ounce	oz
pound	lb
quart	qt
gallon	gal.
inches	in.
foot	ft
yard	yd
mile	mi.
square inch	sq in.
square foot	sq ft
cubic inch	cu in.
cubic foot	cu ft

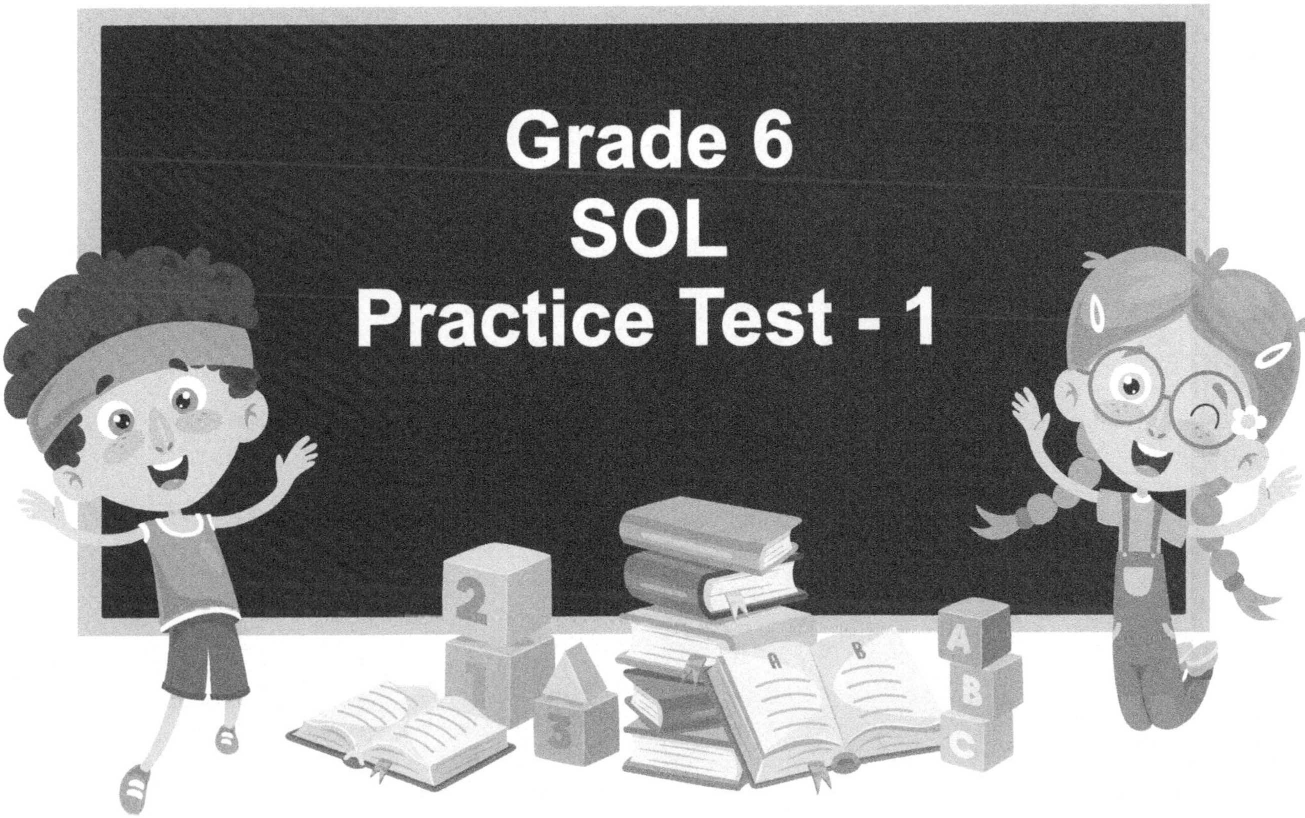
Grade 6
SOL
Practice Test - 1

1. Carla has $50 to spend at the book stores. She buys books which cost $ 7.25 and pens for $1.50 each. What information is needed to find the amount of money that Carla will have left after purchasing at the store ?

(A) The number of pencils she bought
(B) The number of books she bought
(C) The number of books and pens she bought
(D) The number of school supplies she bought

2. Tanya bought fruits for $3.64 , $7.87 and $13.20 , including sales tax. She gave the clerk two $40.00. Which is closest to the amount of change Tanya should receive ?

(A) $15.39 (B) $14.79
(C) $15.29 (D) $16.51

3. Evaluate $\frac{24.01}{0.7}$

(A) 34.03 (B) 342
(C) 3.432 (D) 34.3

4. Rik and Rak collected strawberries in baskets from the farm Rak collected $7\frac{1}{7}$ pounds and Rik $4\frac{3}{7}$ pounds. How many pounds of starwberries were collected by them all together ?

(A) $11\frac{4}{7}$ (B) $10\frac{6}{7}$

(C) $11\frac{3}{7}$ (D) $10\frac{2}{7}$

5. Pamela buys 2 loafs of bread at $2.99 per loaf, 4 pounds of bananas at $0.55 per pound, and a dozen of eggs at $1.85 per dozen. Which is closest to the total cost of the items in her shopping cart ?

(A) $5.39 (B) $10.00
(C) $10.50 (D) $7.04

6. Evaluate 26.9 ÷ 0.5 = ?

(A) 26.4 (B) 27.1
(C) 12.401 (D) 53.8

7. Evaluate $\frac{1}{2} \div 3\frac{1}{4} = ?$

(A) $\frac{2}{13}$ (B) $3\frac{3}{4}$

(C) $5\frac{1}{6}$ (D) $6\frac{1}{2}$

8. What is the greatest common factor of

56 , 72

(A) 4 (B) 32

(C) 8 (D) 504

9. Ken repaired a number of vehicles at his garage this month. The data is given in the table below.

ITEM	Number of vehicles re paired	Cost per Item
Car	25	$ 20.99
Truck	5	$ 55.99
Bus	11	$ 25.99

Based on the data in the table above,Which could be used to determine the total amount of Ken made before tax ?

(A) Multiply the cost per item by the number of items, then multiply the product

(B) Divide the cost per item by the number of items, then multiply the quotients

(C) Divide the cost per item by the number of items, then add

(D) Multiply the cost per item by the number of items, then add the products

10. What is the percentage equivalent to $\frac{17}{20}$

(A) $\frac{17}{20}\%$ (B) $17\frac{1}{5}\%$

(C) 85000 % (D) 85%

11. Which of the below is the composite number?

(A) 31 (B) 7

(C) 39 (D) 113

12. Which of the below represents the closest angle measure of 72^0 from the below ?

(A) 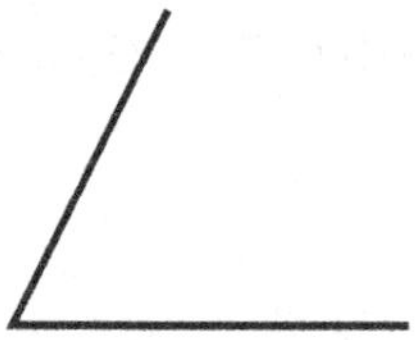

(B) 

(C) 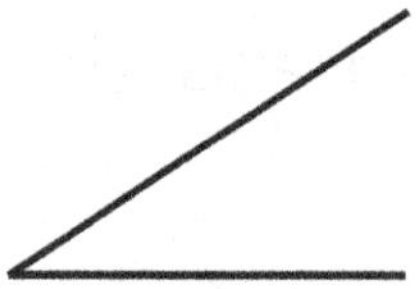

(D) 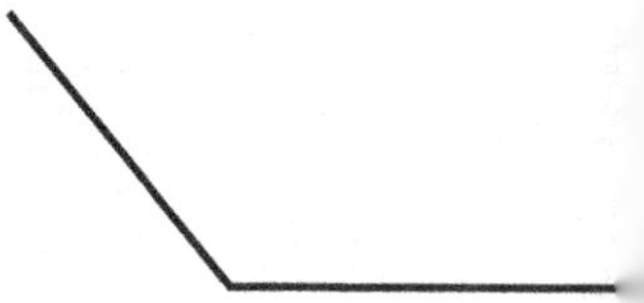

13. Which of the below statements is true ?

(A) $29 > 59$ (B) $-7 > 7$

(C) $-9 > 9$ (D) $-201 < 121$

14. A prime number is ?
    (A) Always an even number
    (B) Has only two factors one and itself
    (C) Always an odd number
    (D) A composite number

15. Which decimals are ordered from least to greatest ?

    (A) 7.63, 7.574, 7.745, 7.7
    (B) 7.7, 7.63, 7.574, 7.745
    (C) 7.745, 7.63, 7.574, 7.7
    (D) 7.574, 7.63 , 7.7, 7.745

16. Caryns needs $4\frac{3}{4}$ cup of floor to make a cake for her daughters Birthday party. She only has $2\frac{3}{5}$ cups of flour .How much floor does she need to make the cake?

    (A) $2\frac{3}{20}$
    (B) $5\frac{1}{20}$
    (C) $2\frac{1}{20}$
    (D) $2\frac{7}{20}$

17. How many degree Celsius is equivalent to 41° F ?

    (A) 4° C
    (B) 6.11° C
    (C) 5° C
    (D) 4.44° C

18. Which of the below is closest to the area of a School Bus ?

(A) 360000 square feet (B) 360 square feet

(C) 30 square feet (D) 36000 feet

19. A wheel has a diameter of 22 meters. Which is closest to the circumference of this wheel ?

(A) 22π m (B) 24π m

(C) 49π m (D) 28π m

20. How many quarts are equivalent to 2 gallons ?

(A) 3 quarts (B) 4 quarts

(C) 6 quarts (D) 8 quarts

21. What is the probability of getting an even number when a six-sided dice is rolled ?

(A) $\frac{1}{2}$ (B) $\frac{1}{3}$

(C) $\frac{1}{6}$ (D) $\frac{4}{6}$

22. The time taken (in seconds) by a group of students in finishing a race is given below. Find the mean time taken by the students to complete the race.

18, 18, 15, 17, 14, 15, 14, 12, 19, 16

(A) 16.8 (B) 17.05

(C) 15.8 (C) 18.05

23. Age of CEO's when assuming office in the year 2018 - 19 are given below :

Stem	Leaf
3	4
4	1 2 2 6 7 7 7
5	0 0 1 5 5 5 5 5
6	5

Key :
1|3 = 13

Which of the following is a *true* conclusion based on the data in the stem-and-leaf plot ?

(A) Most CEO's age is between 25 - 30 years.
(B) Three CEO's have an age of 45.
(C) Lowest age of CEO is 41 years.
(D) Five CEO's have an age of 55.

24. The heights of 5 friends Amy , Beth , Tanya , Cathy and Zoya are given below. What is the mean height of these girls ?
50 inches , 56 inches , 54 inches , 48 inches and 60 inches.

(A) 54.06 inches

(B) 54.56 inches

(B) 53.6 inches

(D) 53.06 inches

25. Store A sells various sizes of shoes. The store sales of this week is given below

6 , 6.5 , 6.5 , 7 , 7.5 , 8.5 , 8.5 , 9 , 9.5 , 10 , 11

Which of the below options represents the data given above

(A)

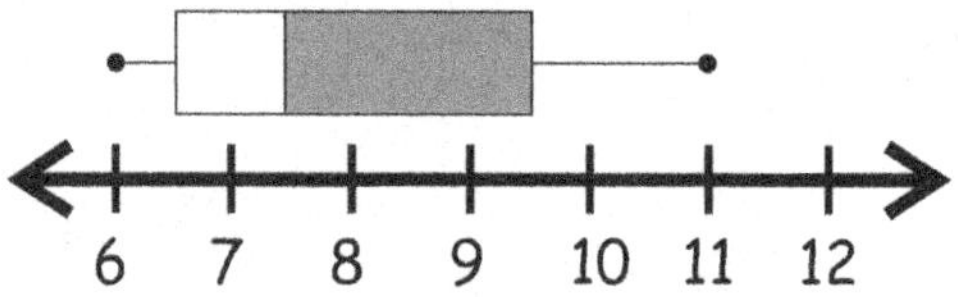

(B)

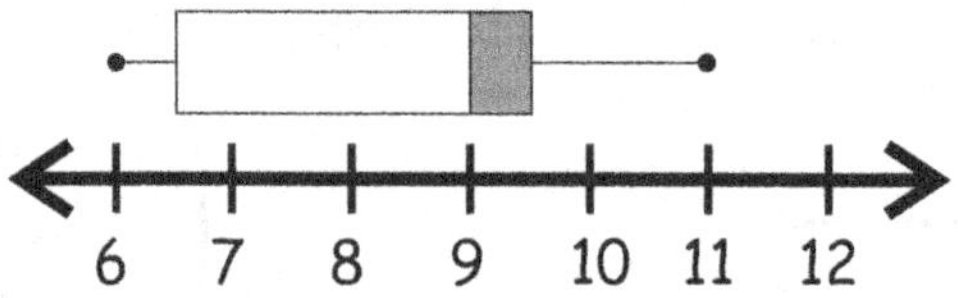

(C)

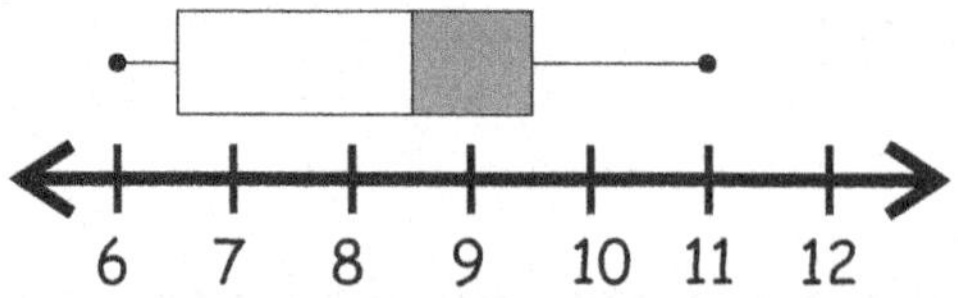

(D)

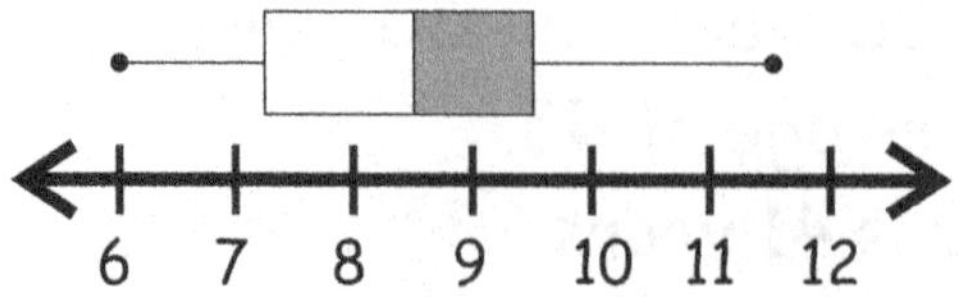

26. Dan is getting ready for a dinner party at his friends home. He has few options to choose as below.

Dan's options

Shirts	Pants	Shoes
Blue	Grey	Sneakers
Green	Black	Slip-Ons

Based on the above data, which of the following shows all of Dan's possible clothing combinations of 1 shirt, 1 pant and 1 type of shoes for the dinner party ?

(A)
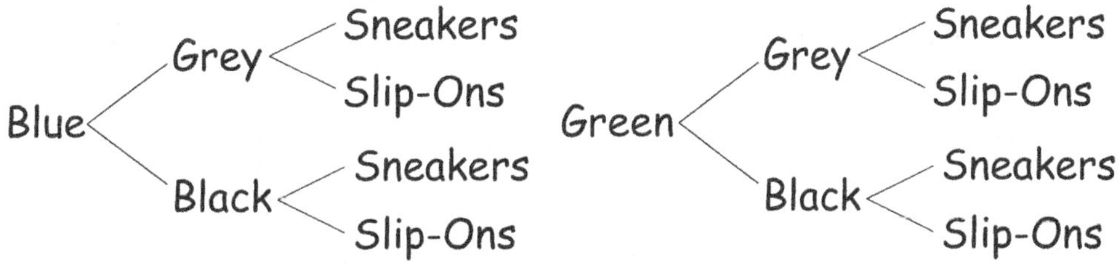

(B)
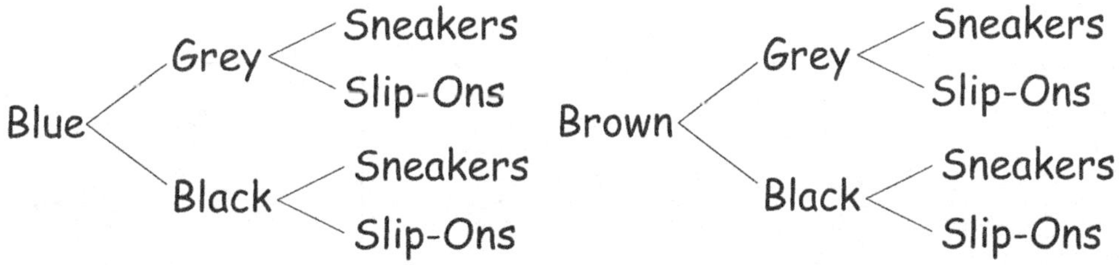

(C)
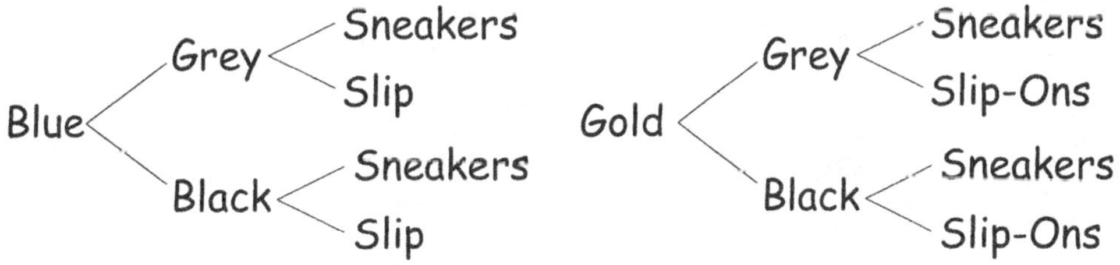

(D)
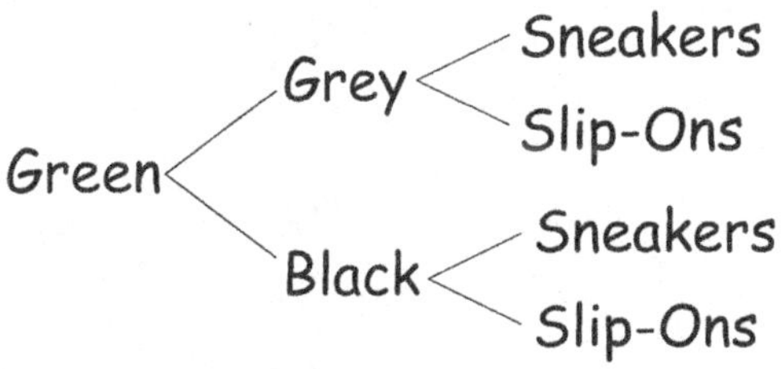

27. The below number sentence is true for what value of "x" ?

$$x + 10 = 59$$

(A) -4 (B) 49

(C) 69 (D) $5\frac{9}{10}$

28. Which word best describes the circled number

$$-17\,ⓝ = -85$$

(A) Co-efficient (B) Variable

(C) Term (D) Equation

29. Based on the below equation. Find the value of Z

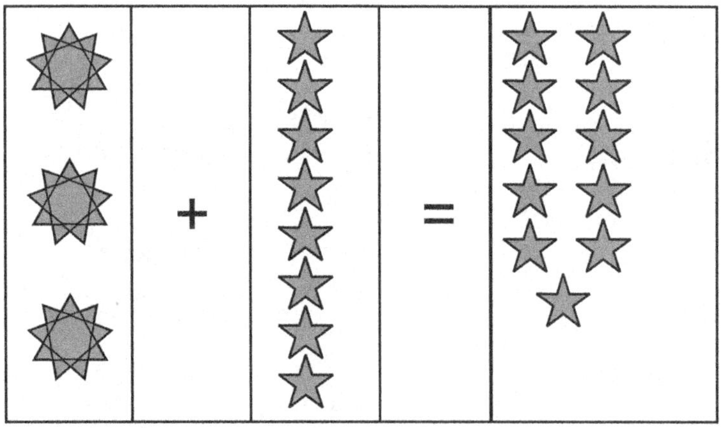

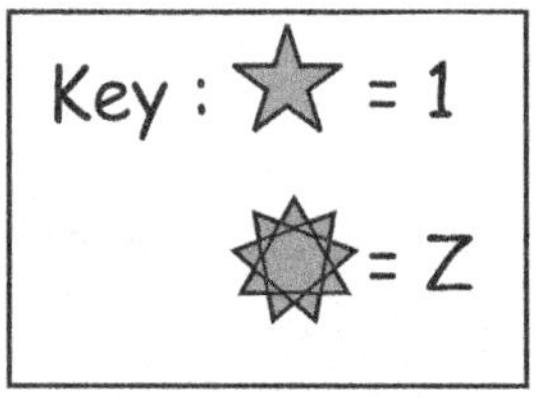

(A) 3 (B) 0

(C) 6 (D) 1

30. Find the solution for the below equation ?

$$\frac{n}{14} = 11$$

(A) 154 (B) 25

(C) $\frac{11}{14}$ (D) -3

31. Find the 6th number in the Arithmetic pattern

192 , 293, 394,...

(A) 495 (B) 786

(C) 697 (D) 897

32. Which of the below is not a perfect square ?

(A) 121 (B) 144

(C) 360 (D) 49

33. Evaluate the below :
If $2^0 = 1$, $2^1 = 2$, $2^2 = 4$, $2^3 = 8$ then $2.2^3.2^2 = ?$

(A) 2^4 (B) 2^{-6}

(C) 2^5 (D) 2^6

34. What is the 13th number in the below pattern ?

123 123 12

(A) 2 (B) 3

(C) 1 (D) 0

35. What number sentence is modeled by the shapes shown ?

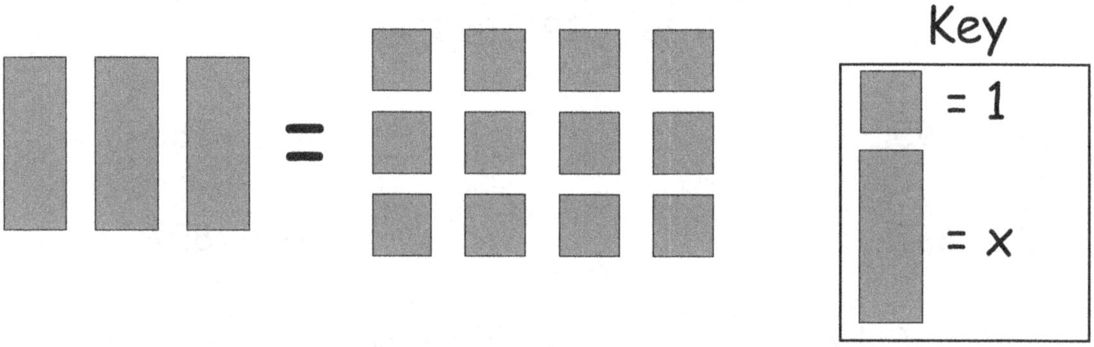

(A) x + 12 = 12 (B) 3x = 12

(C) 12x = 3 (D) x = 9

36. What is the 7th term in the pattern given below ?

10 , 15 , 25 , 40 ....

(A) 170 (B) 210

(C) 105 (D) 65

37. The below pattern is based on a rule. Which of the below options defines the rule ?

8 ,48 , 288, 1728, ...

(A) Divide the previous number by 2.
(B) Multiply the previous number by 6.
(C) Multiply the previous number by 40
(D) Add 40 to the previous number.

38. The variable in the below expression is ?

$$9x - 17 = 302$$

(A) -17
(B) 9
(C) -
(D) x

39. Find the area of a circle with radius as 14 inches.(Approximately)

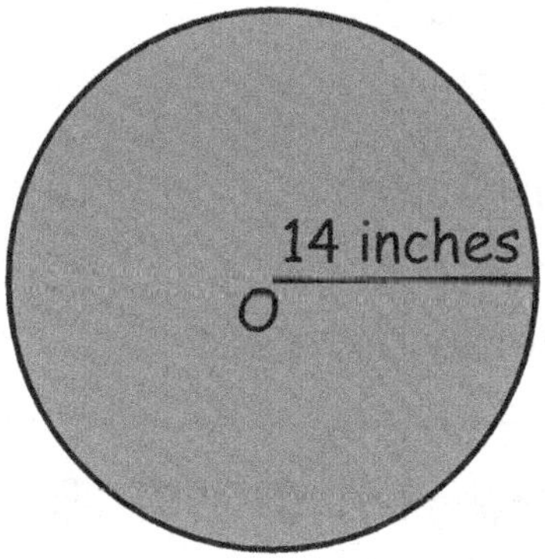

(A) 616 inches
(B) 216 sq inches
(C) 784 sq inches
(D) 616 sq inches

40. Find the least common multiple of the below

21 , 14

(A) 42 (B) 14

(C) 294 (D) 7

41. Evaluate $1\frac{1}{2} \times \frac{2}{3} \times \frac{1}{4}$

(A) $\frac{1}{2}$ (B) $3\frac{5}{12}$

(C) $\frac{1}{4}$ (D) $\frac{3}{12}$

42. Which of the below represents all the factors of 64 ?

(A) 1 , 2 , 8 , 16 , 36 , 64
(B) 1 , 2 , 4 , 8 , 16 , 32 , 64
(C) 1 , 2 , 4 , 6 , 16 , 64
(D) 1 , 2 , 4 , 6 , 8 , 32 , 16 , 64

43. Find the median of the below data set

40 , 42 , 46 , 46 , 48 , 51 , 55 , 58 , 63 , 70

(A) 49.5 (B) 62

(C) 55.4 (D) 62.5

44. Find the range of the below data set

5.76 , 1.69 , 4.93 , 1.52 , 1.92 , 1.8 , 2.58 , 4.09 , 1.44

(A) 4.38 (B) 5.47

(C) 4.32 (D) 3.2

45. Find the missing ∠x of the below quadrilateral ABCD

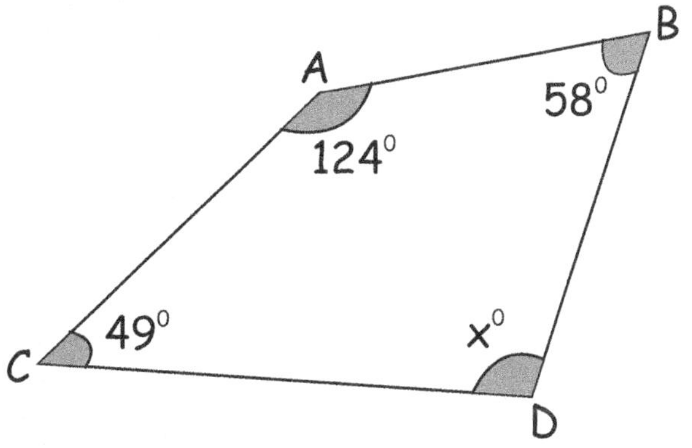

(A) 112^0 (B) 118^0

(C) 129^0 (D) 110^0

46. How many feet equals to 33 yards ?

(A) 99 (B) 19

(C) 15 (D) 3

47. Find the area of the triangle ABC given below

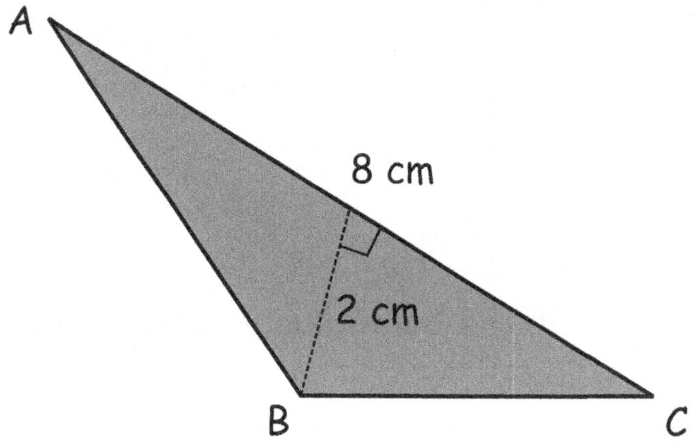

(A) 4 cm^2 (B) 8 cm

(C) 8 cm^2 (D) 16 cm^2

48. To find which of the below the middle value of the data set is needed ? (Assuming data set has odd number of values)

(A) Mode (B) Range

(C) Mean (D) Median

49. Find the net of the 3D figure given below

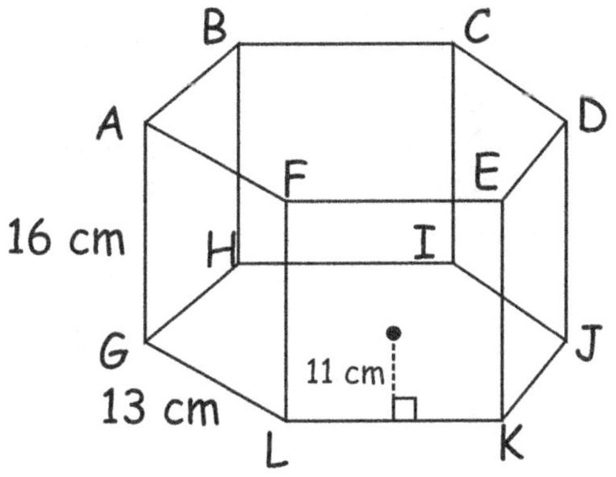

(A) 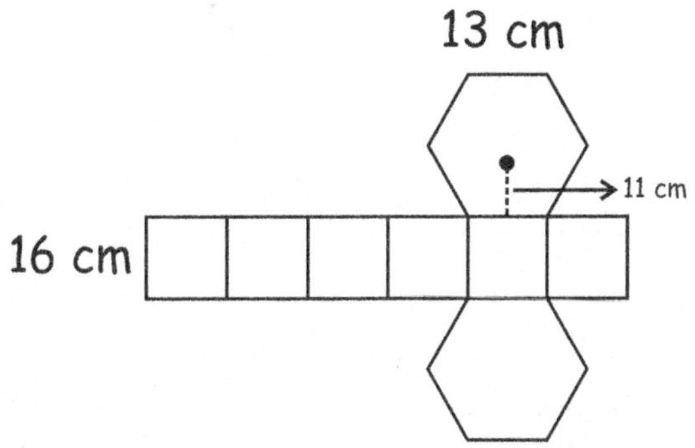

(B) 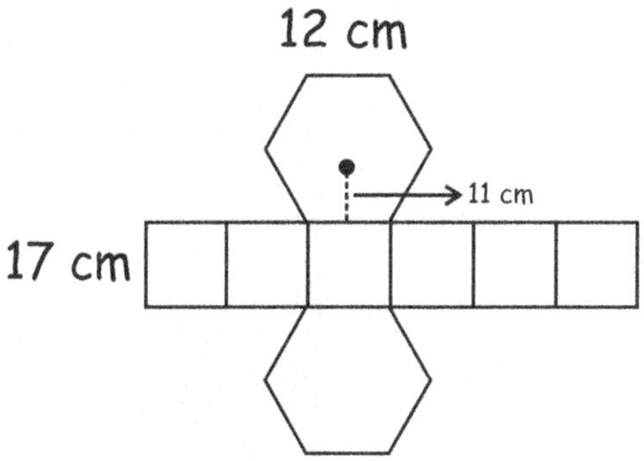

(C) 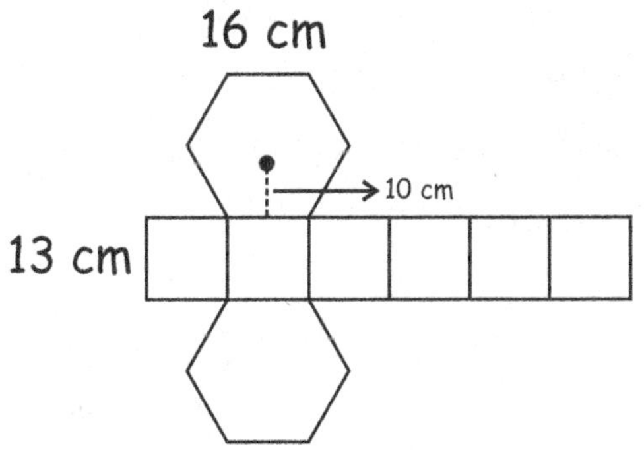

(D) 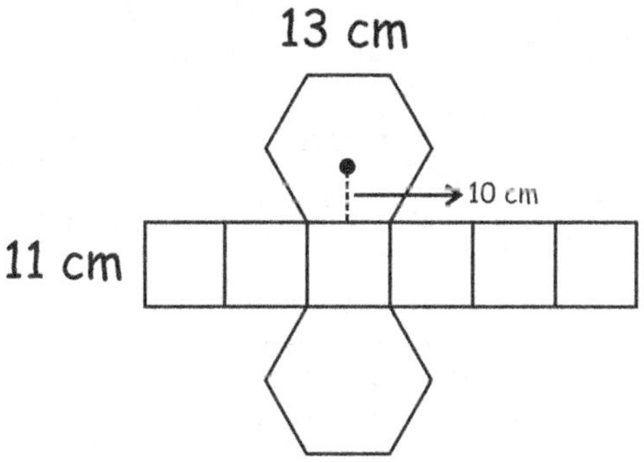

50. The below bar graph shows the production of paper (in million tons) from three different companies A , B , C in last five years.

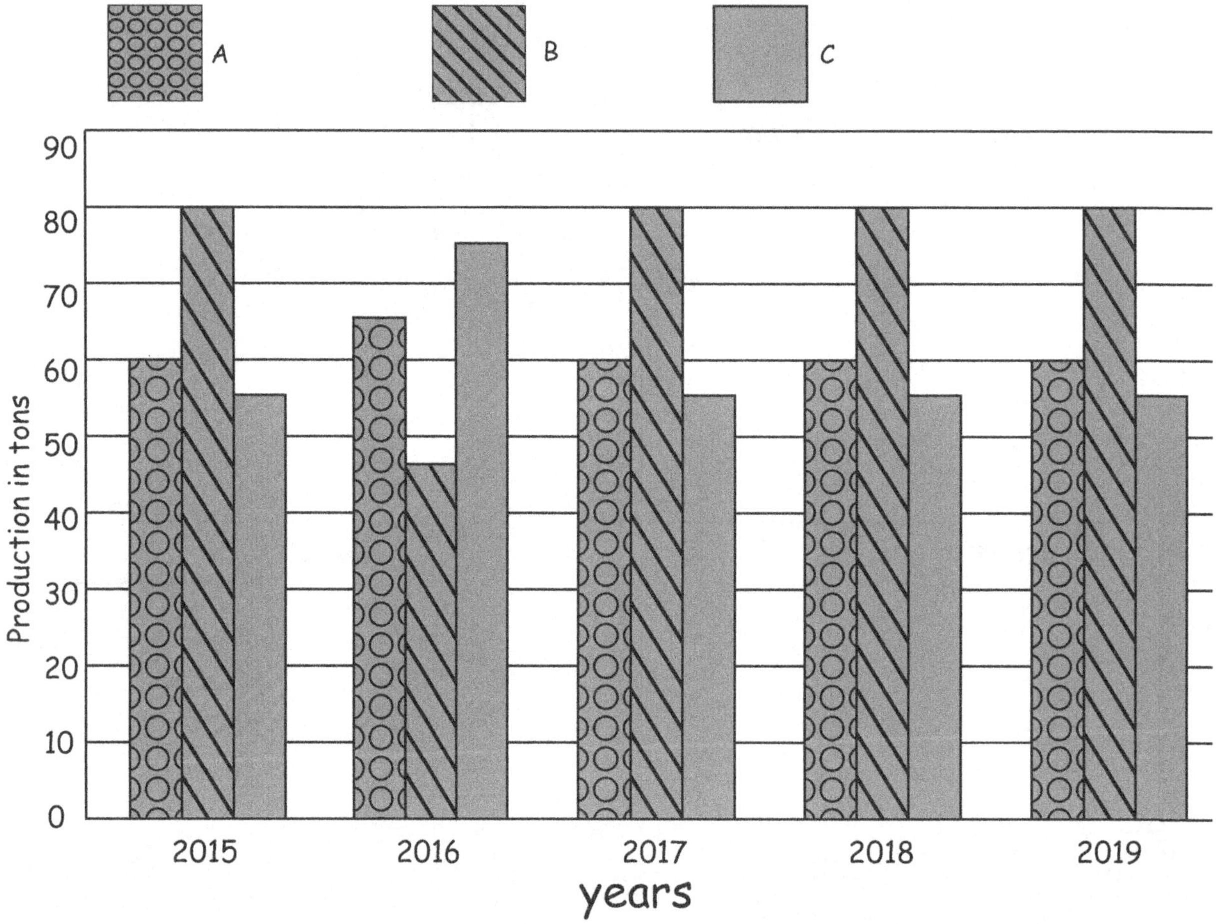

(I) What is the difference between the production of company A in 2016 and company B in 2015 (million tons) ?

(A) 35
(B) 31
(C) 34
(D) 15

(II) What is the ratio of the average production company C from 2016 and 2018 to the average production in company A in the same period ?

(A) 35 : 46
(B) 27 : 26
(C) 26 : 25
(D) 23 : 25

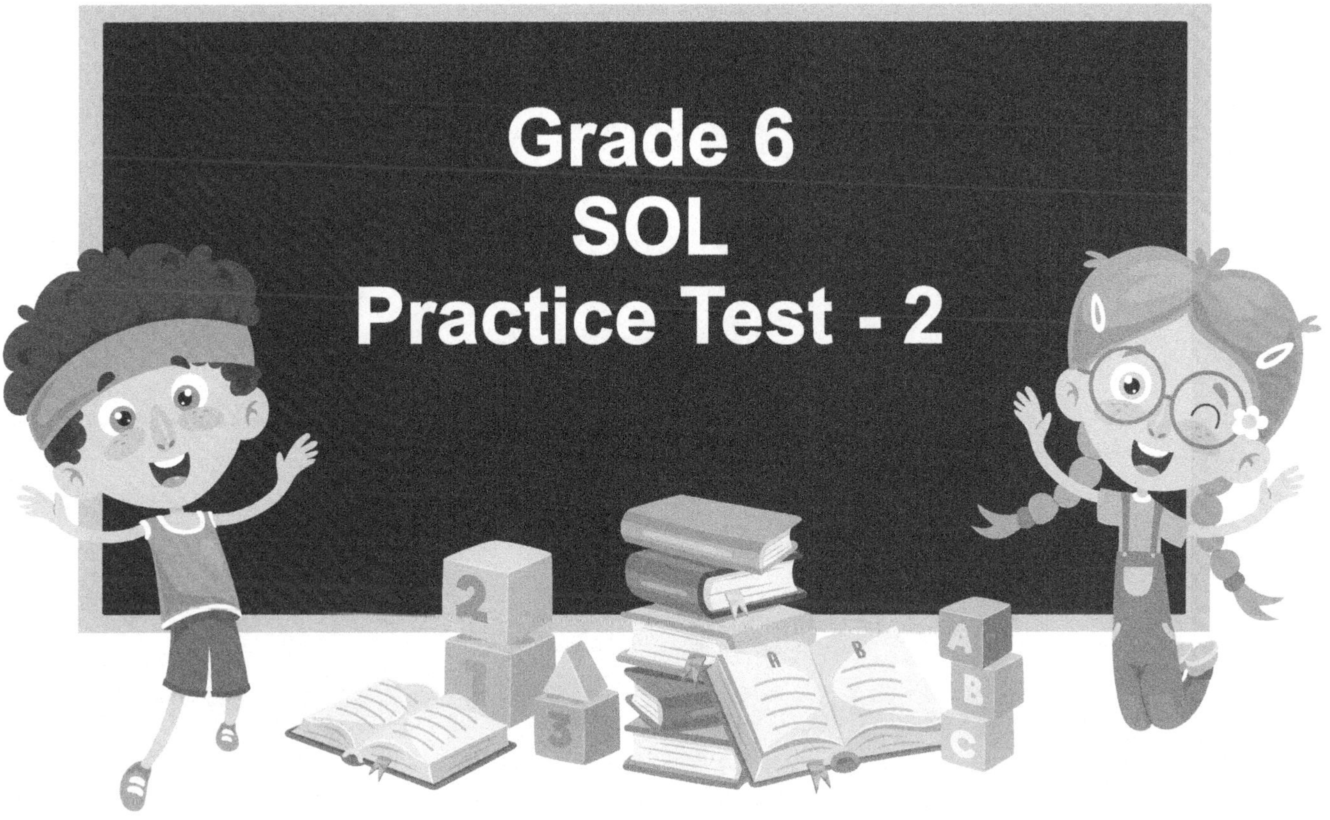
Grade 6
SOL
Practice Test - 2

1. Macy sold bracelets for $2.50 and earrings for $3.75 at the fair. To calculate the amount of money Macy got by selling bracelets and earrings she needs which information ?

(A) The number of earrings sold
(B) The number of bracelets sold
(C) The number of earrings and bracelets sold
(D) The number of necklaces sold

2. Cathy bought snacks for $5.64 , $17.07, and $5.50 including sales tax from three different stores. She has $30.00. Which is closest to the amount Cathy has after spending for snacks ?

(A) $1.80 (B) $2.07

(C) $1.15 (D) $1.71

3. Evaluate 45.36 ÷ 0.7 = ?

(A) 64.80 (B) 64.08

(C) 6408 (D) 640.08

4. Jack and Jill collected many Apples in baskets from the apple farm. Jack collected $11\frac{3}{8}$ pounds and Jill $9\frac{7}{8}$ pounds. How many more pounds of apples were collected by Jack ?

(A) $3\frac{1}{4}$ (B) $1\frac{3}{4}$

(C) $8\frac{1}{4}$ (D) $1\frac{1}{2}$

5. Tom buys 3 milk cans at \$4.07 per can , 3 pounds of oranges at \$0.55 per pound , and 1 cheese pack at \$5.35. Which is closest to the total cost of the items ?

(A) \$9.97 (B) \$19.20

(C) \$20.00 (D) \$29.25

6. Evaluate $1.1 \div 27.5 = ?$

(A) 0.04 (B) 5.04
(C) 20.9 (D) 25

7. Evaluate $\frac{1}{7} \div \frac{9}{5} = ?$

(A) $\frac{2}{5}$ (B) $\frac{9}{35}$

(C) $\frac{5}{63}$ (D) $5\frac{11}{12}$

8. What is the greatest Common factor of

33 , 27

(A) 15 (B) 297

(C) 9 (D) 3

9. Stacy is selling baked goods to rise funds for her school STEM.

ITEM	Number sold	Cost per Item
Sugar cookies	32	\$ 0.50
Chocolate chip cookies	49	\$ 0.60
Brownies	50	\$ 1

Based on the data in the table above , Which could be used to determine the total amount Stacy raised at her school bake sale before tax ?

(A) Multiply the cost per item by the number of items, then add the products

(B) Divide the cost per item by the number of items, then multiply the quotients

(C) Divide the cost per item by the number of items, then add

(D) Multiply the cost per item by the number of items, then multiply the product

10. What is the percentage equivalent to $\frac{1}{33}$

(A) $30\frac{10}{33}$ %

(B) $1\frac{33}{100}$

(C) $\frac{1}{33}$ %

(D) $3\frac{1}{33}$ %

11. Which of the below is the composite number ?

(A) 67

(B) 87

(C) 89

(D) 23

12. Which of the below represents the closest angle measure of 50^0

(A)

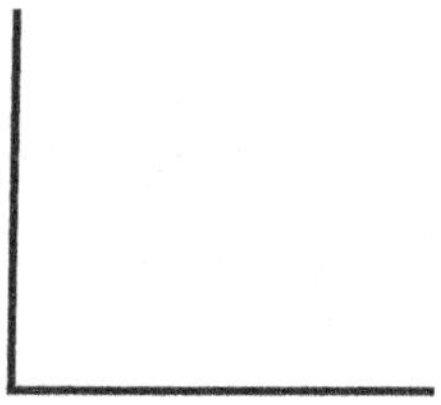

(B)

(C)

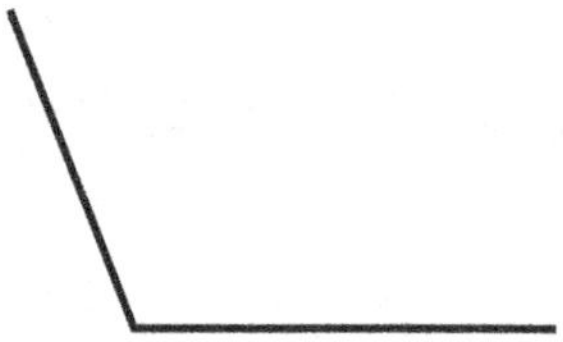

(D)

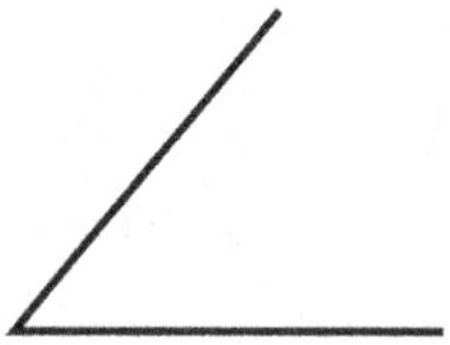

13. Which of the below statements is true ?

(A) $901 < 619$

(B) $-4 > 5$

(C) $-11 < 11$

(D) $-51 > 99$

14. A composite number is ?
    (A) Units digit is always 7
    (B) Units digit is odd number
    (C) Always divisible by 5
    (D) Not a prime number

15. Which decimals are ordered from least to greatest ?

    (A) 0.39 , 0.956 , 0.5 , 0.3 , 0.971
    (B) 0.3 , 0.39 , 0.5 , 0.956 , 0.971
    (C) 0.971 , 0.3 , 0.956 , 0.39 , 0.5
    (D) 0.39 , 0.5 , 0.971 , 0.956 , 0.3

16. Jack ran $3\frac{1}{5}$ miles on Saturday and $2\frac{1}{4}$ mile on Sunday.

    How many miles did he ran this weekend ?

    (A) $5\frac{9}{20}$
    (B) $8\frac{17}{60}$
    (C) $1\frac{47}{60}$
    (D) $7\frac{33}{40}$

17. How many lbs are equivalent to 88 ounces ?

    (A) 5 lbs
    (B) 6.5 lbs
    (C) 4.5 lbs
    (D) 5.5 lbs

18. Which are the closest L X W X H of a Land Rover vehicle ?

(A) 4970 X 1990 X 1888 mm
(B) 497 X 180 X 180 mm
(C) 8900 X 6500 X 7000 mm
(D) 9123 X 5174 X 2001 ft

19. A wheel has a diameter of 20 meters. Which is closest to the circumference of this wheel ?

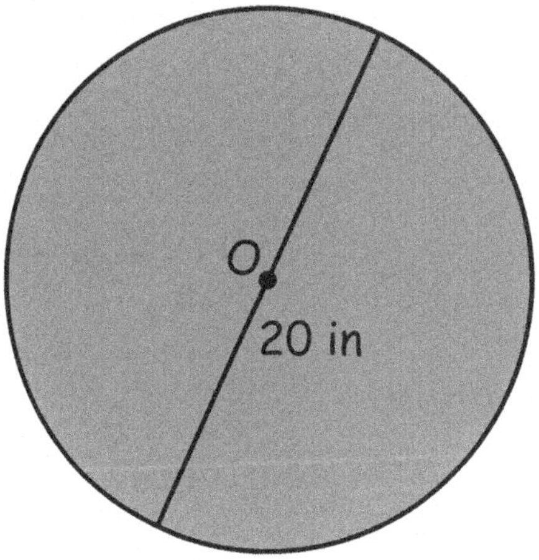

(A) 16∏ in
(B) 22∏ in
(C) 20∏ in
(D) 23∏ in

20. How many quarts are equivalent to 11 pints ?

(A) 3 quarts
(B) 5.5 quarts
(C) 6 quarts
(D) 12 quarts

21. A box contains 5 blue stars,6 gold stars and 9 red stars. One star is randomly picked from the box. What is the probability that is a gold star ?

(A) $\frac{5}{20}$ (B) $\frac{9}{20}$

(C) $\frac{1}{20}$ (D) $\frac{6}{20}$

22. The time taken (in seconds) by a group of students in finishing a race is given below. Find the mean time taken by the students to complete the race.

18 , 18 , 15 , 17 , 14 , 15 , 14 , 12 , 19 , 16

(A) 16.8 (B) 17.05

(C) 15.8 (C) 18.05

23. Average Students Heights

Stem	Leaf
4	9
5	3 5 5 5 5 9 9 9
6	0 0 2 3 4 4
7	1

Key :
1|3 = 1.3

Which of the following is a *true* conclusion based on the data in the stem-and-leaf plot?

(A) Four students have an average height of 5.5 feet.
(B) Three students have average height of 6 feet.
(C) Minimum average height of students was 4.5 feet.
(D) The maximum average student height was 7.5 feet.

24. Grade 3 students played fall games over the weekend. Number of hits in a round of hacky sack game are plotted as below.

5 , 5 , 6 , 6 , 7 , 8 , 12 , 12

(A)

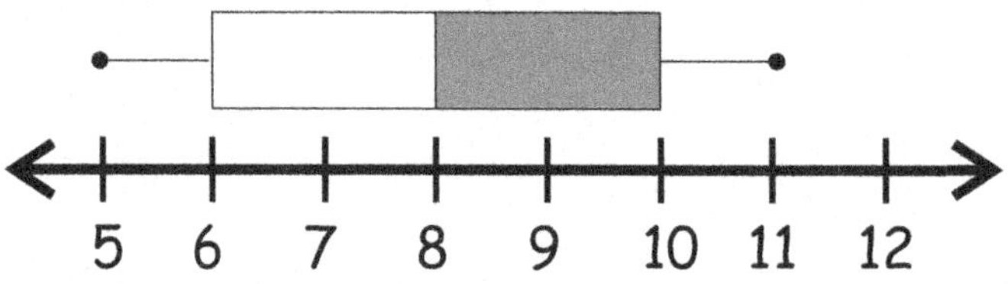

(B)

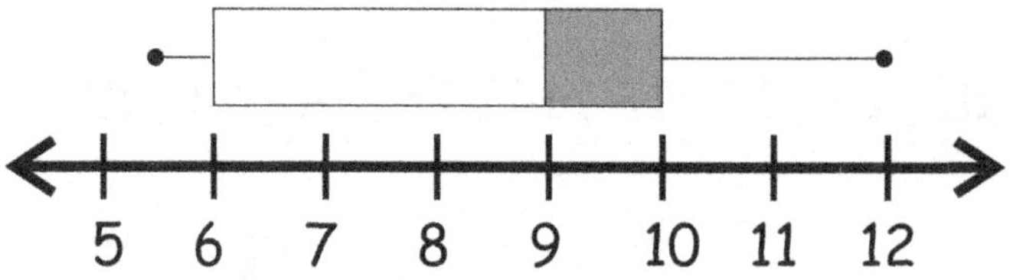

(C)

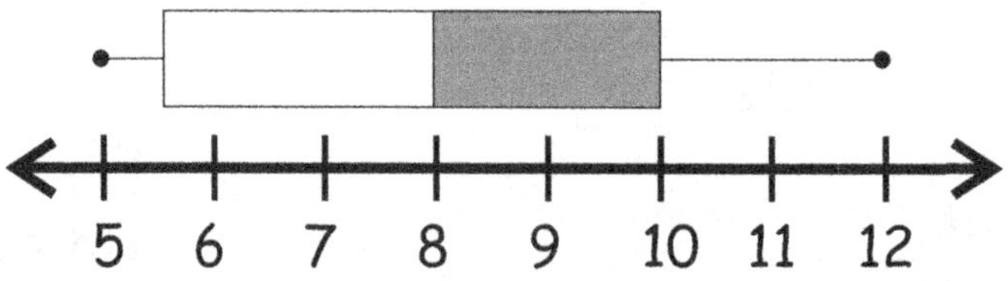

(D)

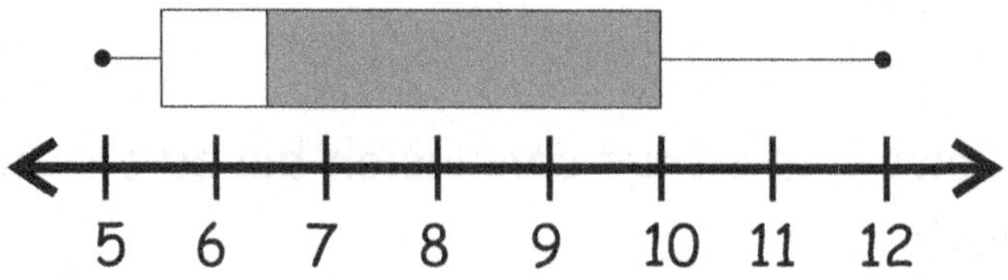

25. Nancy is getting ready for her friends graduation party.

Nancys options

Dress	shoes	hand bag
Gold	Black	Gold
Brown	White	White

Based on the chart, which of the following shows all of Andy's possible clothing combinations of 1 color of shirt, 1 type of pants and 1 type of pants ?

(A)
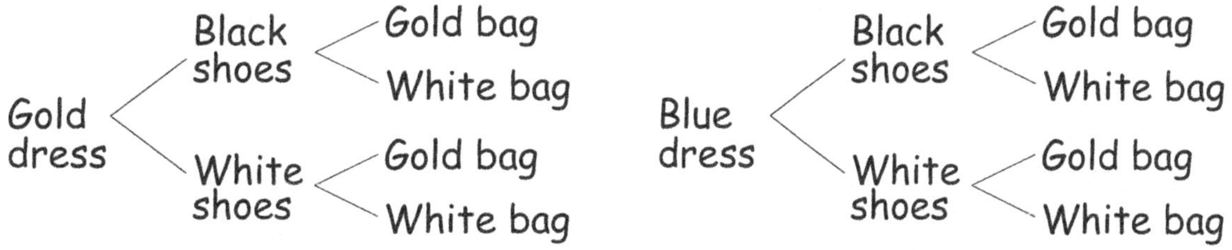

(B)
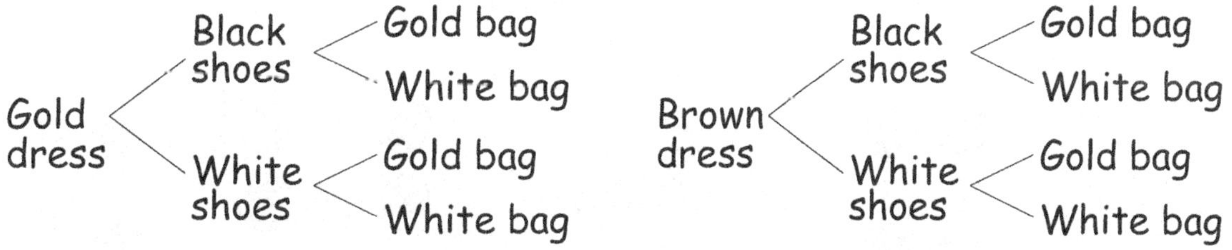

(C)
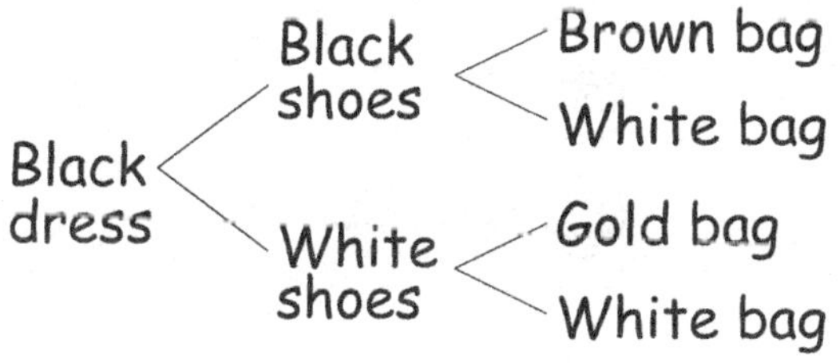

(D)
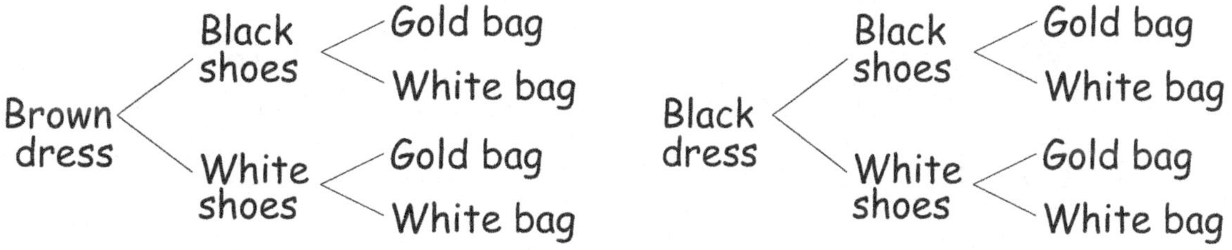

26. The below number sentence is true for what value of "a" ?

$$a + 10 = 16$$

(A) 26　　(B) 6

(C) 4　　(D) $1\frac{3}{5}$

27. which word best describes the circled number

$$\textcircled{9}k = 27$$

(A) Variable　　(B) Equation

(C) Term　　(D) Co-efficient

28. Based on the below equation. Find is the value of of Z

$$6Z - 9 = 15$$

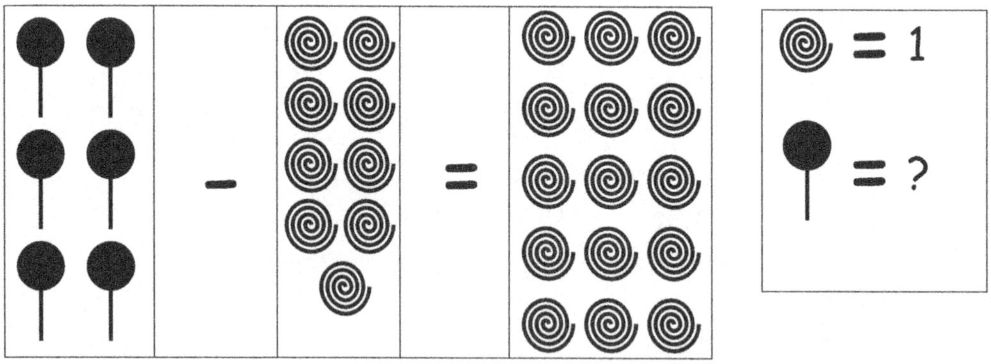

(A) 4　　(B) 2

(C) 1　　(D) 24

29. Find the solution of the below equation ?

$$\frac{19}{13} = \frac{n}{13}$$

(A) 13 (B) -11

(C) -10 (D) 19

30. Find the 7th number in the Arithmetic pattern

123 , 234 , 345 , ______, _______

(A) 789 (B) 489

(C) 567 (D) 689

31. Which of the below is not a perfect square?

(A) 100 (B) 81

(C) 112 (D) 121

32. Evaluate the below :

$$5^2.5^3.5^2$$

(A) 5^5 (B) 5^7

(C) 5^4 (D) 5^9

33. What is the 15th figure in the below pattern?

1234 1234 123

(A) 2 (B) 1

(C) 4 (D) 3

34. What number sentence is modeled by the shapes below ?

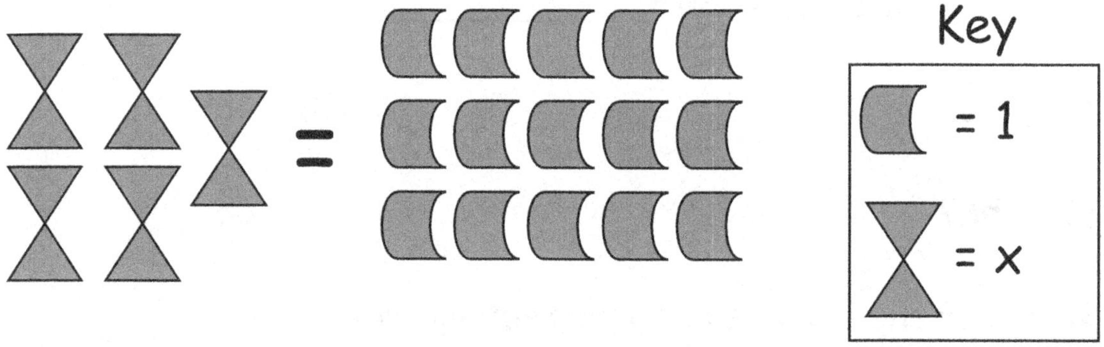

(A) $5x + 4 = 12$ (B) $4x = 6$

(C) $5x = 15$ (D) $x + 15 = 5$

35. What is the 7th term in the pattern given below ?

102 , 165 , 267 , 432 ,..........

(A) 1830 (B) 699

(C) 1131 (D) 1950

36. The below pattern is based on a rule. Which of the below options defines the rule ?

134456 , 19208 , 2744 , 392 ....

(A) Divide the previous number by 7.
(B) Multiply the previous number by 2.
(C) Subtract the previous number by 12306
(D) Add 12306 to the previous number.

37. The variable in the below expression is ?

$6x + 45 = 1230$

(A) 6
(B) x
(C) 45
(D) +

38. Find the least common multiple of the below

27 , 18

(A) 486
(B) 49
(C) 9
(D) 54

39. Evaluate $1\frac{3}{4} \times 2 \times \frac{2}{5} = ?$

(A) $\frac{1}{30}$
(B) 1
(C) $1\frac{2}{5}$
(D) $7\frac{3}{20}$

40. Which of the below represents all the factors of 63 ?

(A) 1 , 3 , 7 , 9 , 21 , 63
(B) 1 , 3 , 7 , 9 , 21 , 36 , 63
(C) 1 , 3 , 9 , 63
(D) 1 , 3 , 7 , 9 , 11 , 23 , 34

41. Find the area of the circle whose radius is 11 mm

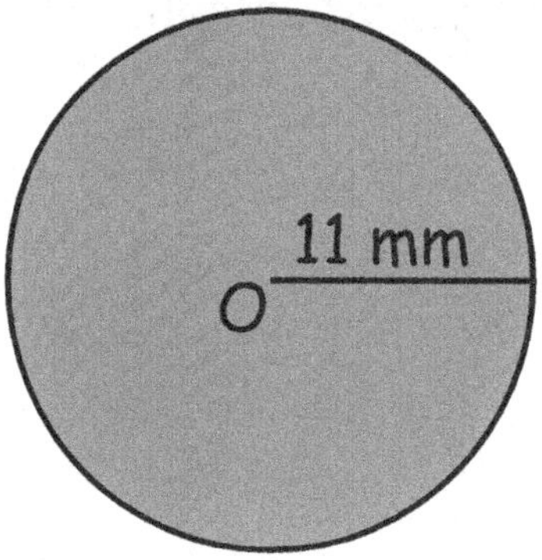

(A) 110 mm^2
(B) 371 mm^2
(C) 380 mm^2
(D) 421 mm^2

42. Find the median of the below data set

53 , 61.5 , 66.5 , 69 , 80.5 , 73 , 77.5 , 82.5 , 82.8

(A) 73.5
(B) 73
(C) 77.7
(D) 69

43. Find the area of the parallelogram ABCD given below

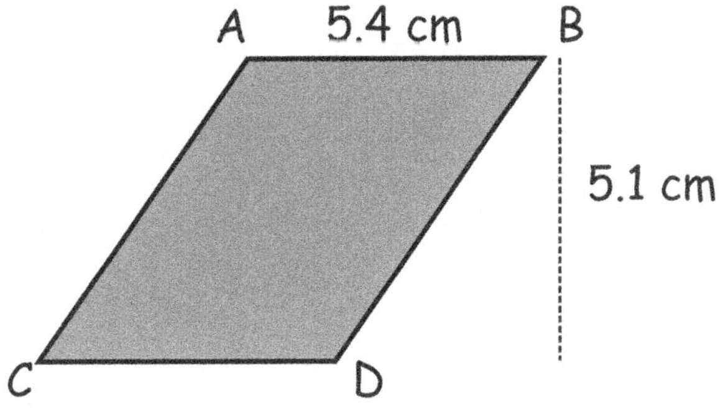

(A) 25.92 cm^2 (B) 27.54 cm^2

(C) 27.05 cm^2 (D) 14.8 cm^2

44. How many inches are equal to 5 yards ?

(A) 120 inches (B) 150 inches

(C) 18 inches (D) 180 inches

45. Find the missing $\lfloor b$ of the below quadrilateral ABCD

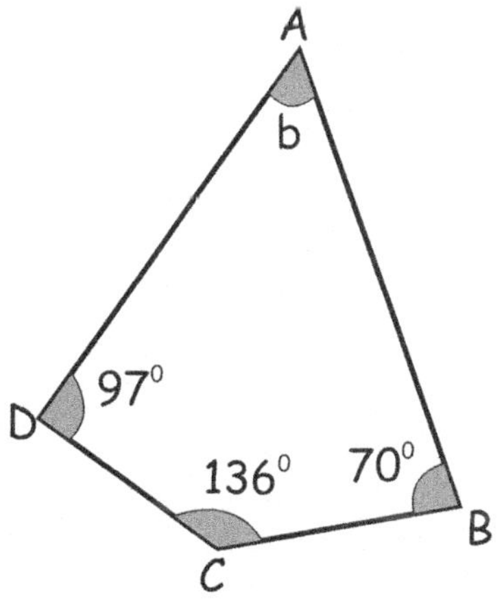

(A) 57^0

(B) 66^0

(C) 64^0

(D) 55^0

46. Find the net of the 3D figure given below

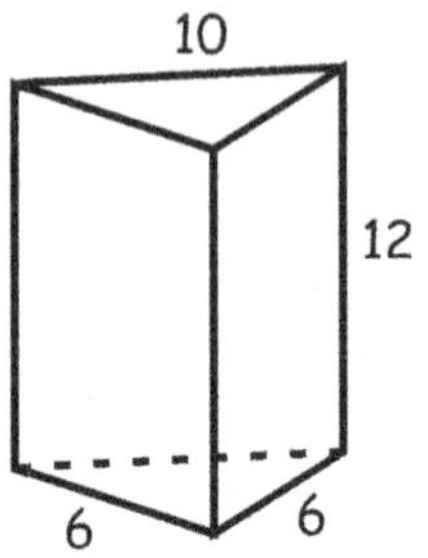

(A)

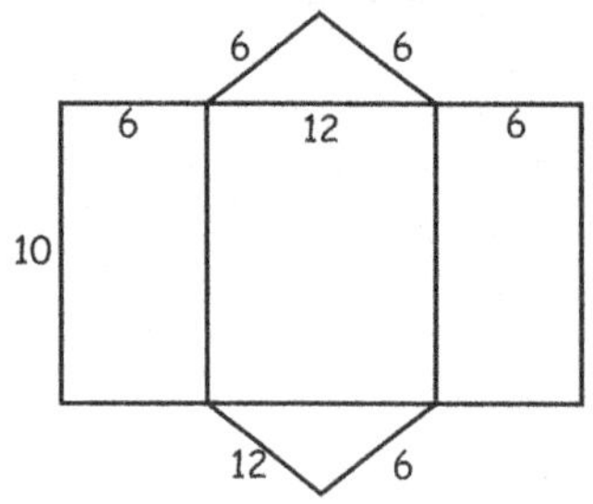

(B)

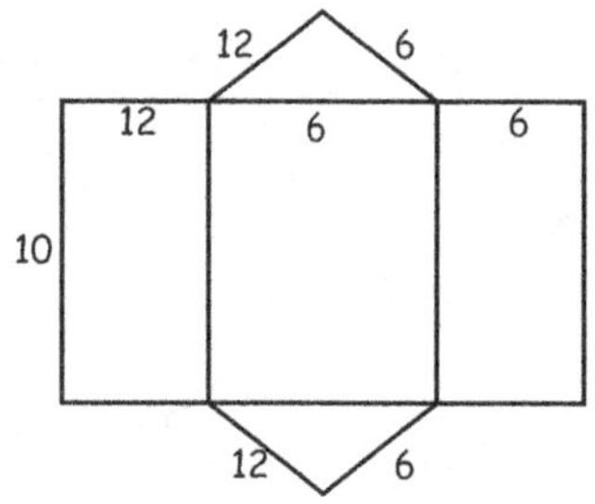

(C)

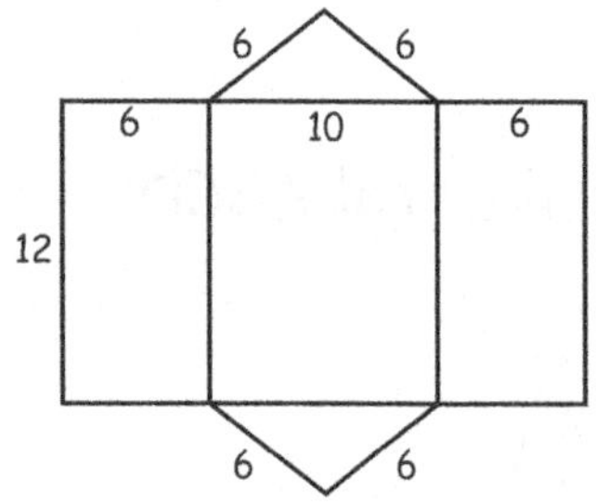

(D)

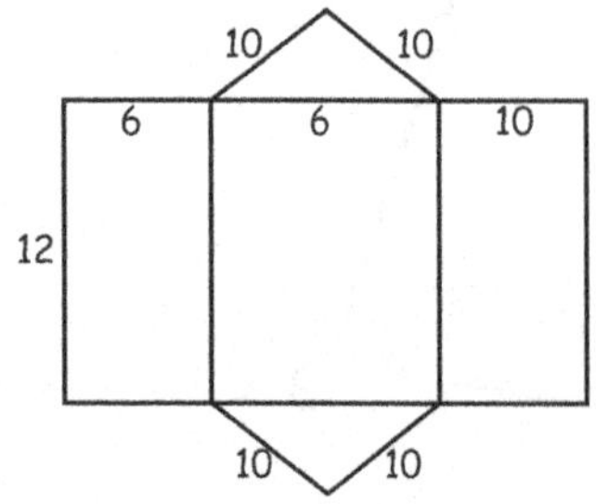

The below bar graph shows the total number of pizzas (in hundreds) sold by 3 different stores in 6 different months in the year 2018.

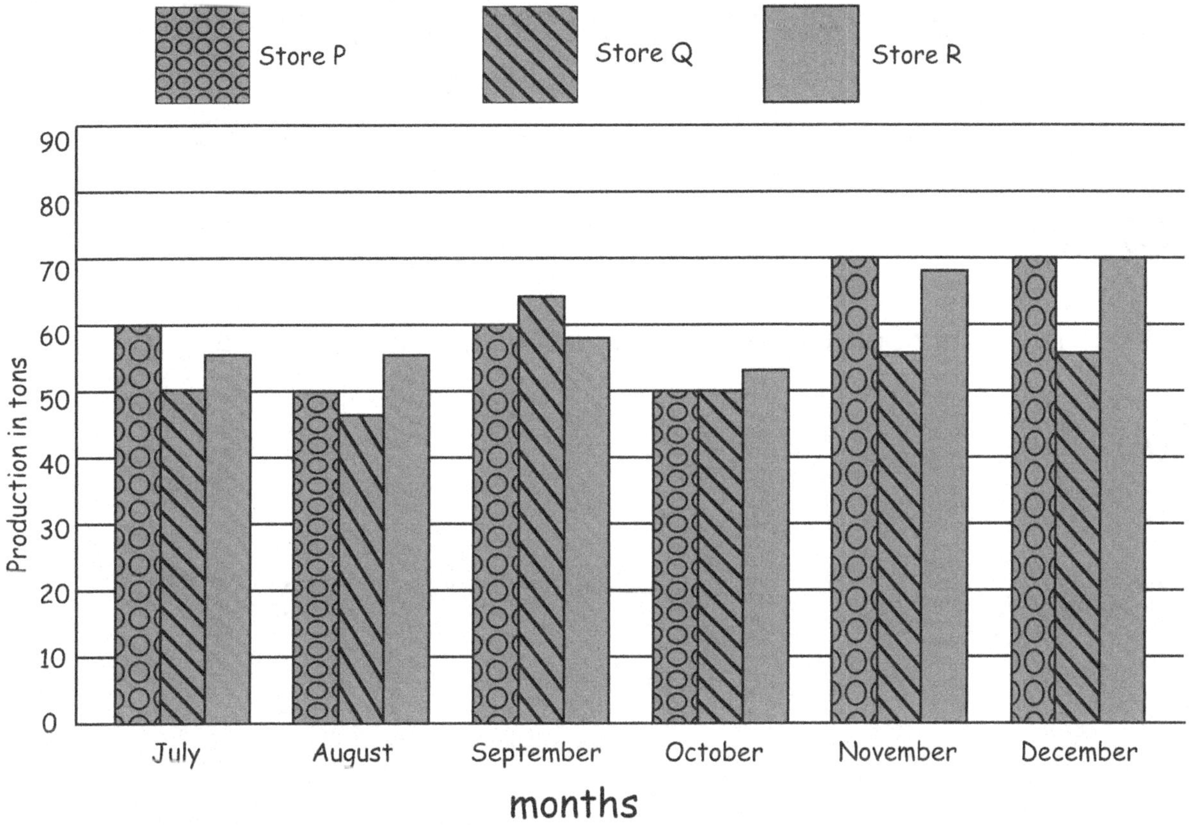

47. Find the average number of pizzas sold by store P in all given months together ?

(A) 6000 (B) 5650

(C) 6100 (D) 62500

48. Find the difference between the total number of pizzas sold by Store Q to that of store R in December ?

(A) 2600 (B) 1200

(C) 3000 (D) 1500

49. Find the range of the below data.

77.9 , 78.6 , 74.1 , 74.1 , 80.3 , 79.3 , 83.8 , 79.6 , 80.3

(A) 6.9 (B) 9.6

(C) 9.7 (D) 7.8

50. What is the probability of getting a multiple of 2 when a dice is rolled ?

(A) $\frac{1}{2}$ (B) $\frac{1}{4}$

(C) $\frac{1}{3}$ (D) $\frac{3}{4}$

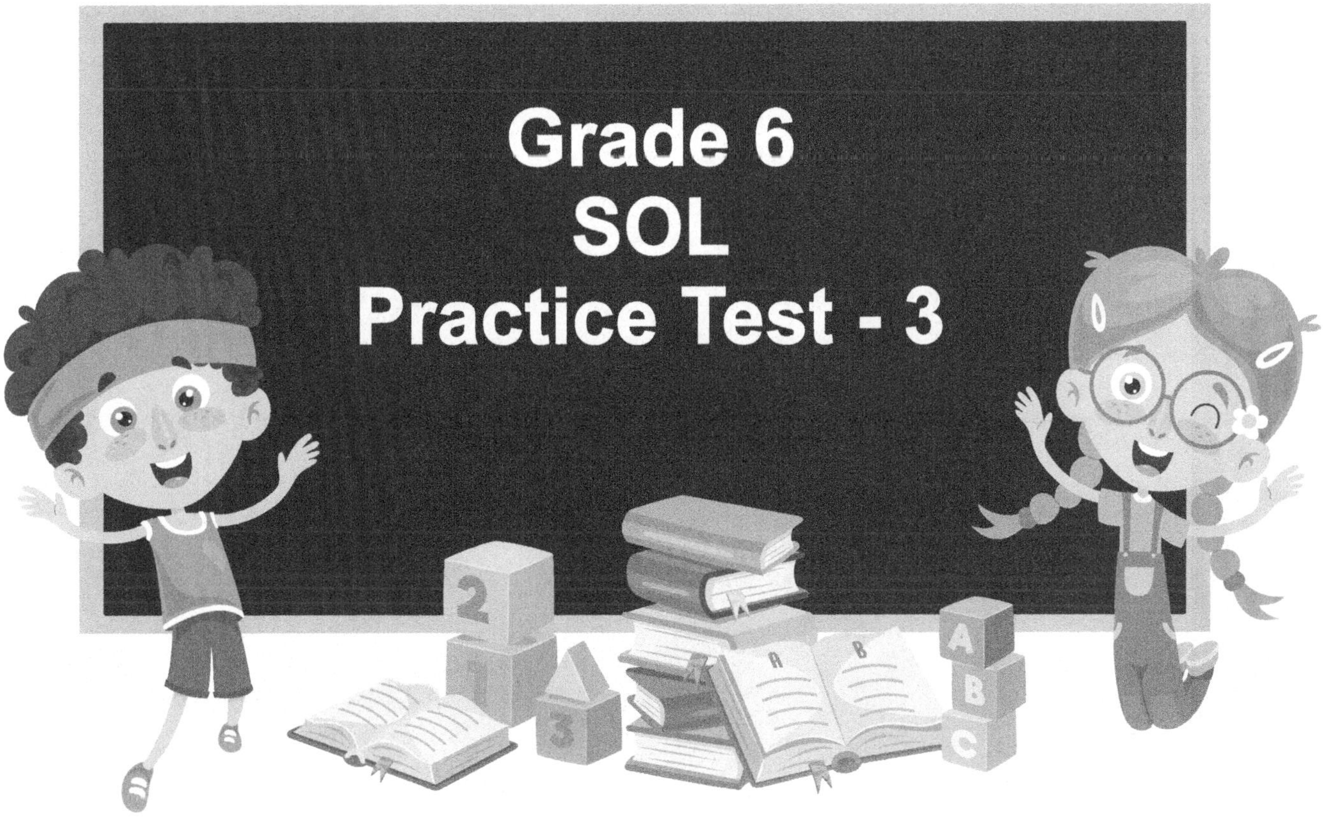
Grade 6
SOL
Practice Test - 3

1. Tara has $250 to spend at the mall. She buys dresses for $ 67.50 each and shoes for $50.50 each. What information is needed to find the amount of money that Tara will have left after purchasing at the mall ?

(A) The number of dresses and shoes that were purchased
(B) The number of dresses she bought
(C) The number of shoes she bought
(D) The amount of tax she paid

2. Amy bought fruit drinks for $10.14, $21.27, and $43.90 including sales tax. She gave the clerk $100.00. Which is closest to the amount Amy has after spending for fruit drinks ?

(A) $24.39 (B) $24.79

(C) $24.69 (D) $26.57

3. Evaluate $13.31 \div 1.1 = ?$

(A) 12.12 (B) 12.1

(C) 131 (D) 1.412

4. Matt and Mike ran $15\frac{3}{4}$ miles and $13\frac{1}{4}$ miles. How many more miles Matt ran than Mike on the trial?

(A) 29 (B) $\frac{21}{4}$

(C) $29\frac{1}{4}$ (D) $2\frac{1}{2}$

5. Amy buys two sheets of stamps for 3.75 each ,six envelopes for $ 0.35 each and a glue bottle for $4.90
Find the total cost of the items that she bought ?

(A) $14.60 (B) $11.55

(C) $14.50 (D) $10

6. Evaluate $5.7 \div 2.5 = ?$

(A) 16.5 (B) 2.28

(C) 29.5 (D) 25.9

7. Evaluate $1\frac{5}{13} \div 1\frac{2}{5} = ?$

(A) $5\frac{4}{13}$ (B) $\frac{1}{9}$

(C) $\frac{90}{91}$ (D) $1\frac{61}{65}$

8. What is the greatest common factor of

65 , 39

(A) 195 (B) 26

(C) 5 (D) 13

9. Jason sold various candles at arts and crafts festival as per the data given below

ITEM	Number of candles sold	Cost per Item
Small	25	$ 3.50
Medium	5	$ 5.25
Large	11	$ 7.00

Based on the data in the table above , Which could be used to determine the total amount of Jason made before tax ?

(A) Multiply the cost per item by the number of items, then multiply the product

(B) Multiply the cost per item by the number of items, then add the products

(C) Divide the cost per item by the number of items, then add

(D) Divide the cost per item by the number of items, then multiply the quotients

10. What is the percentage equivalent to $\frac{27}{50}$

(A) 275%

(B) $\frac{27}{50}$

(C) 54 %

(D) 540%

11. Which of the below is the composite number ?

(A) 807

(B) 449

(C) 661

(D) 937

12. Angle measure closest to 109^{0}

(A)

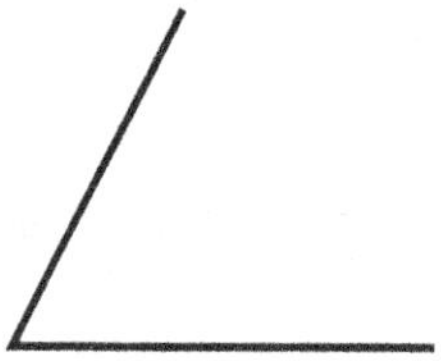

(B)

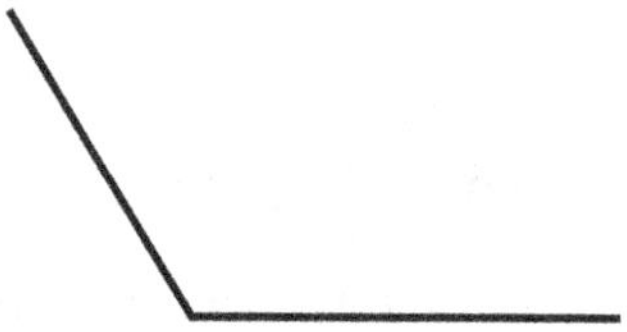

(C)

(D)

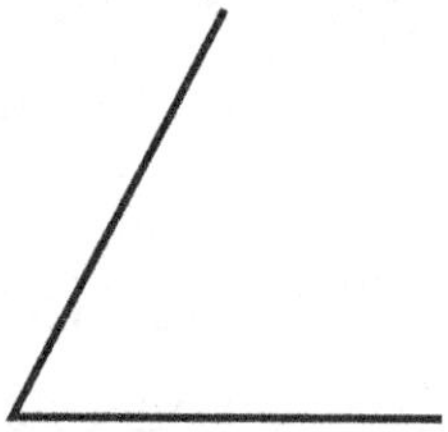

13. Which of the below statements is true ?

(A) $155 < 81$

(B) $-666 < 666$

(C) $-90 < -170$

(D) $-551 > 330$

14. Can a prime number be even ?

(A) Two is the only even prime number.
(B) Prime number is always an odd number
(C) Prime number is composite
(D) None of the above

15. Which decimals are ordered from greatest to least ?

(A) 1.66 , 1.51 , 1.940 , 1.849 , 1.812
(B) 1.849 , 1.812 , 1.940 , 1.66 , 1.51
(C) 1.940 , 1.849 , 1.812 , 1.66 , 1.51
(D) 1.66 , 1.849 , 1.940 ,1.812 , 1.51

16. Rosy wants to make a scarf of length $4\frac{4}{5}$ feet. She already made $3\frac{3}{5}$ feet of scarf. How much more length of scarf does she need to make ?

(A) $4\frac{12}{35}$
(B) $1\frac{1}{5}$
(C) $1\frac{3}{5}$
(D) $8\frac{2}{5}$

17. How many feet are equivalent to 84 inches ?

(A) 9 ft
(B) 6.5 ft
(C) 6.5ft
(D) 7 ft

18. Which is closest to the area of a full bed ?

(A) 8500 square feet
(B) 54 square inches
(C) 54 X 75 square feet
(D) 75 feet

19. A wheel has a diameter of 18 feet Which is closest to the circumference of this wheel ?

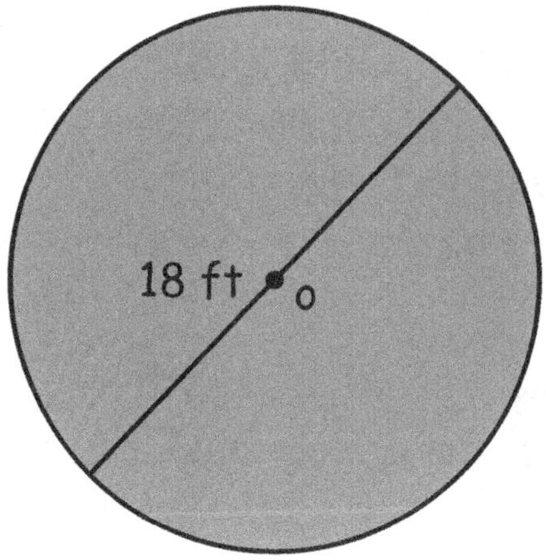

(A) 18∏ ft
(B) 42∏ ft
(C) 21∏ ft
(D) 20∏ ft

20. Which of the below equivalent to 12 yards ?

(A) 432 inches
(B) 48 inches
(C) 468 inches
(D) 422 inches

21. A box contains 5 vowels , 21 consonants. One alphabet is randomly picked from the box. What is the probability that is a vowel ?

(A) $\frac{21}{26}$ (B) $\frac{5}{26}$

(C) $\frac{5}{21}$ (D) $\frac{1}{5}$

22. Find the mean of first 10 prime numbers.

(A) 11.8 (B) 12.09

(C) 11.09 (D) 12.9

23. Age of program managers when assuming office

Stem	Leaf
2	8
3	3 7
4	2 5 5 5 5 7
5	4
6	0 1 2 4 4

Key :
1|3 = 13

Which of the following is a *true* conclusion based on the data in the stem-and-leaf plot ?

(A) Five people have age of 50 years.
(B) Maximum age of program managers is 60 years.
(C) Four program managers have an age of 45 years.
(D) 21 years is the age of the youngest program manager.

24. Below data represents the goals made in a soccer tournament

3 , 4 , 5 , 5 , 6 , 6 , 9

(A)

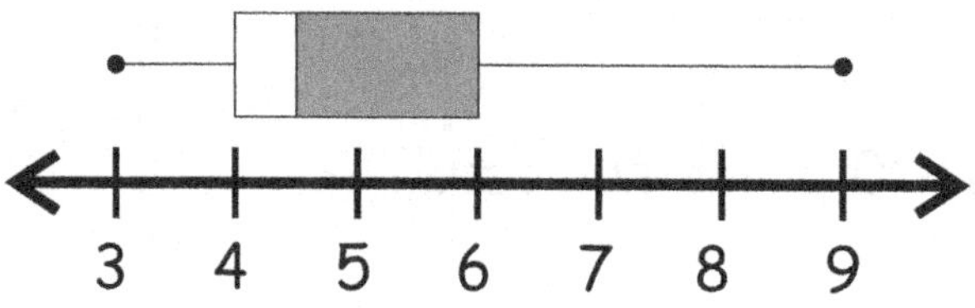

(B)

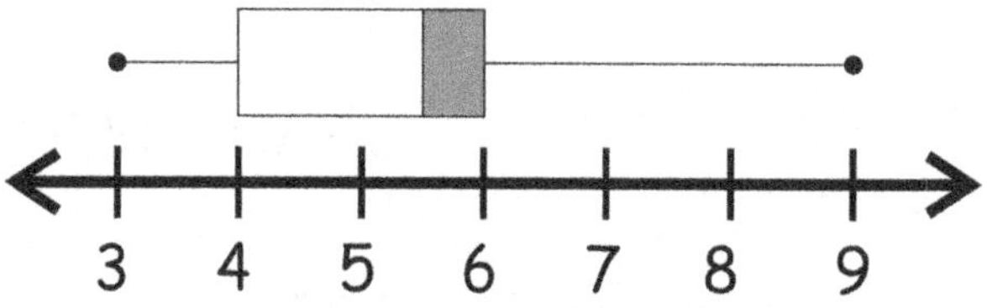

(C)

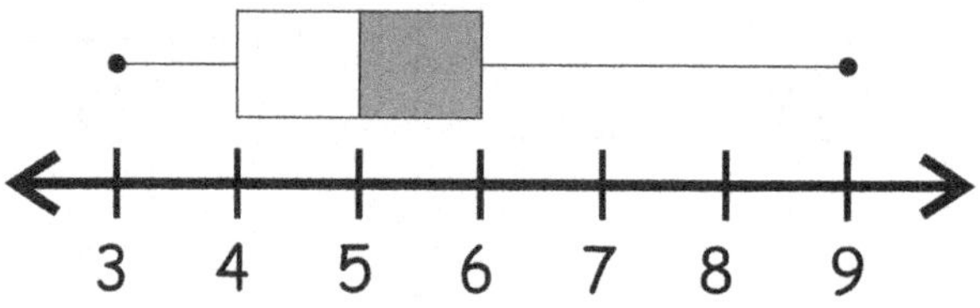

(D)

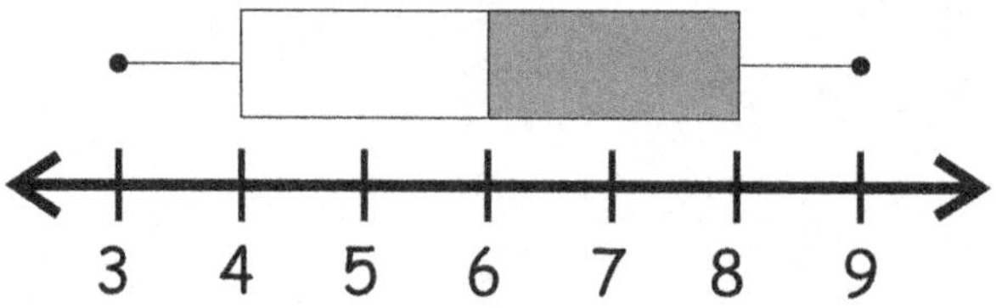

25. Yummy ice cream store sells 2 flavors of ice creams with few choices of toppings.

Rosy's options

Ice cream flavor	Topping 1	Topping 2
Strawberry	Sprinkles	Hot fudge
Mint	Nuts	Caramel

Based on the chart, which of the following shows all of Rosy's possible choices of picking with 2 toppings.

(A)

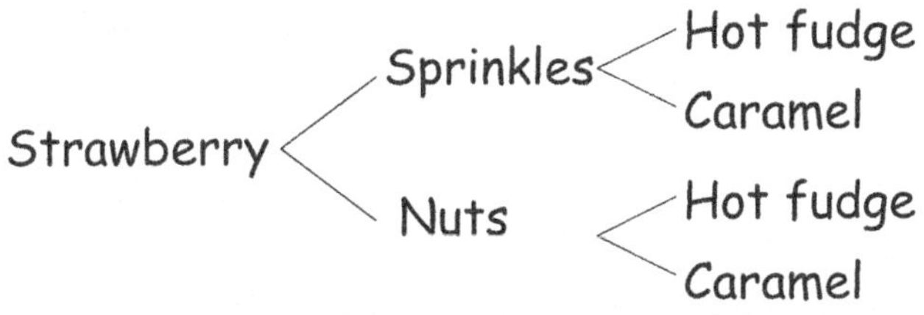

(B)

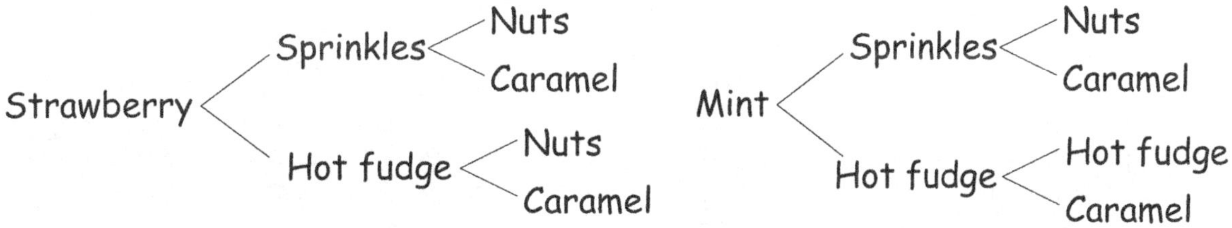

(C)

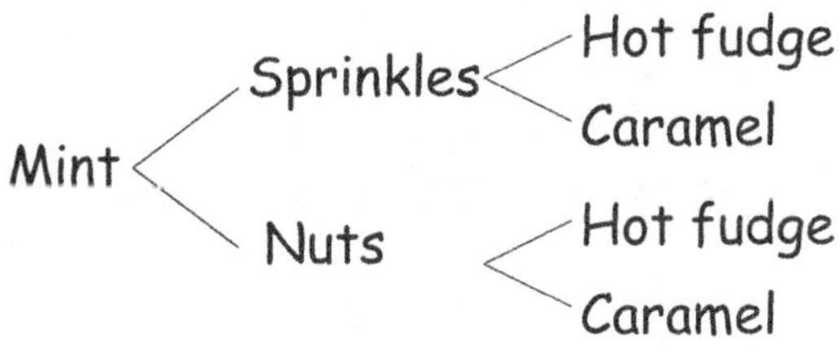

(D)

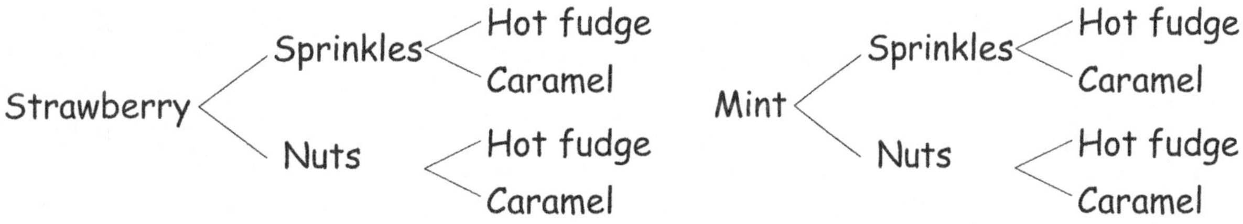

26. The below number sentences true for what value of y

$$n - 10 = -13$$

(A) -3 (B) $-1\frac{3}{10}$

(C) 5 (D) -23

27. which word best describes the circled number

$$7x = \textcircled{-98}$$

(A) Variable (B) Co-Efficient

(C) Term (D) Constant

28. Based on the below equation. Find is the value of of Z

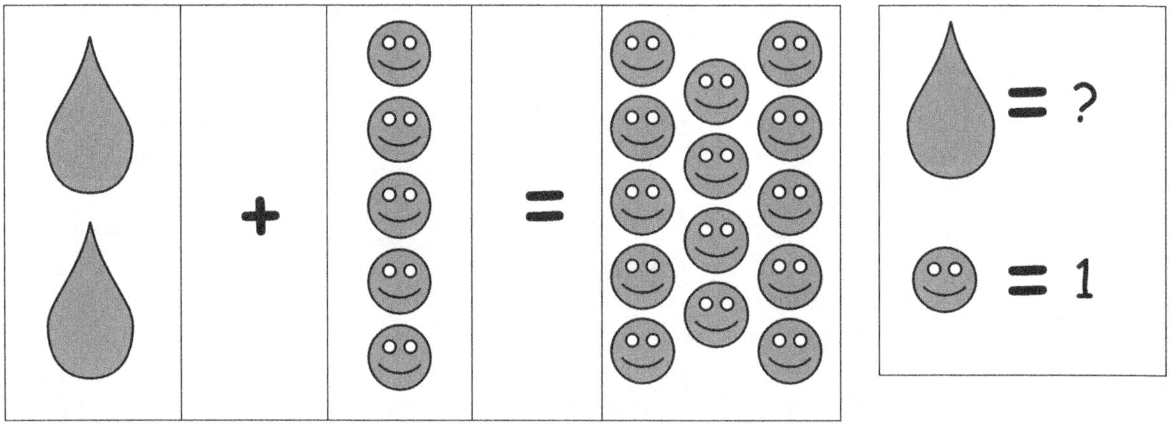

(A) 4.5 (B) 4

(C) 19 (D) 8.5

29. Find the solution of the below equation ?

$$\frac{x}{20} = -7$$

(A) $\frac{-7}{20}$ (B) -27

(C) -140 (D) -13

30. Find the 9th number in the Arithmetic pattern

111 , 222 , 333 , 444

(A) 888 (B) 666

(C) 999 (D) 555

31. Which of the below is not a perfect square?

(A) 99 (B) 169

(C) 625 (D) 16

32. Evaluate the below :

$$4^0.4^3.4^1$$

(A) 4^3 (B) 4^6

(C) 4^5 (D) 4^4

33. What is the 10th figure in the below pattern ?

12121...

(A) 1 (B) 2

(C) 0 (D) 12

34. What number sentence is modeled by the shapes shown ?

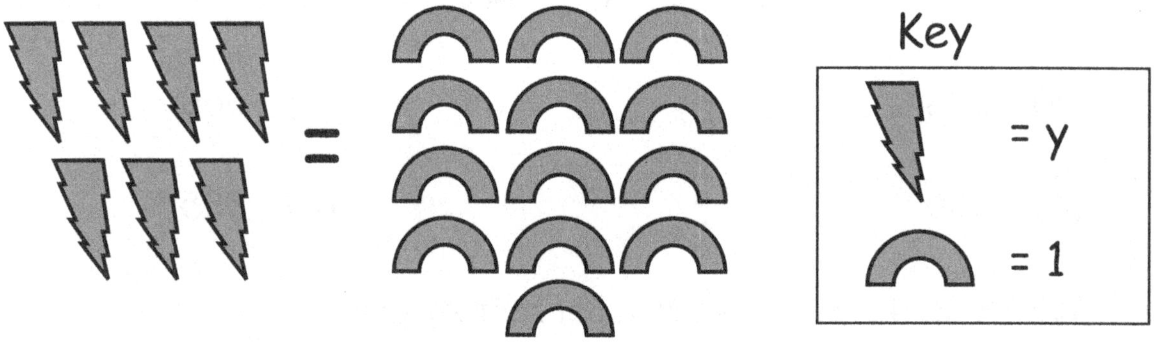

(A) $7y = 13$ (B) $14 + y = 13$

(C) $y - 14 = 4$ (D) $13y = 14$

35. What is the 7th term in the pattern given below ?

1398 , 2262 , 3660 , 5922,.....

(A) 25086 (B) 9582

(C) 15504 (D) 16824

36. The below pattern is based on a rule. Which of the below options defines the rule?

6250 , 1250 , 250 , 50 , ...

(A) Add 40 to the previous number.
(B) Multiply the previous number by 5
(C) Subtract the previous number by 100
(D) Divide the previous number by 5

37. The variable in the below expression is ?

99Y - 909 = 99

(A) 909 (B) 99

(C) -909 (D) Y

38. Which of the below represents all the factors of 98

(A) 1 , 14 , 25 , 49 , 98 (B) 1 , 7 , 14 , 49 , 98

(C) 1 , 2 , 7 , 14 , 49 , 98 (D) 1 , 2 , 7 , 49 , 98

39. Find the area of the circle with a radius of 10 cm

(A) 340 cm^2 (B) 373 cm^2

(C) 314 cm^2 (D) 445 cm^2

40. Find the net of the 3D figure given below

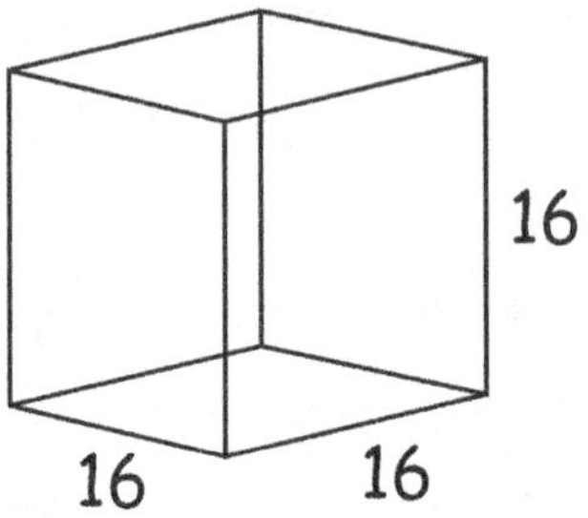

(A)

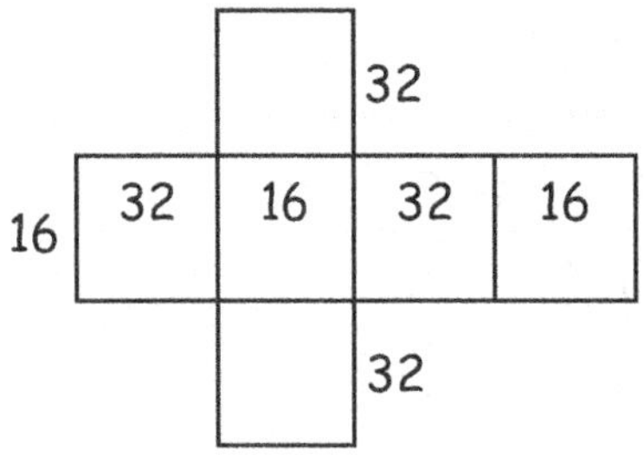

(B)

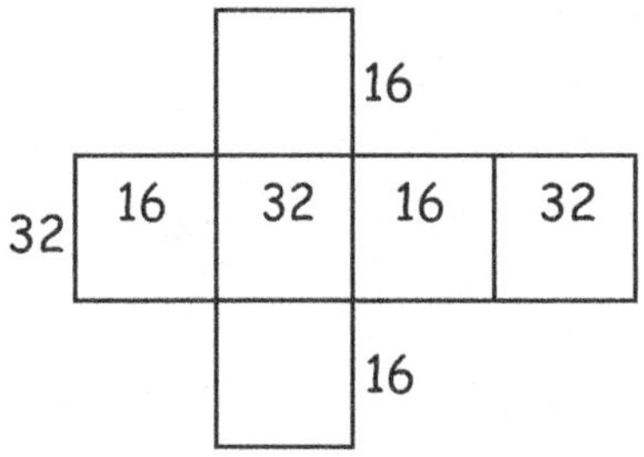

(C)

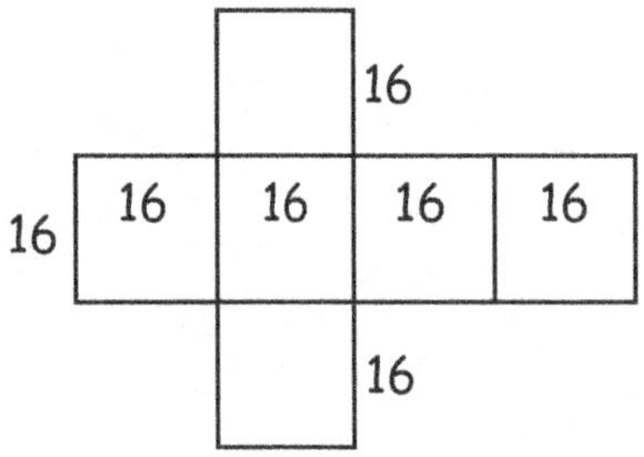

(D)

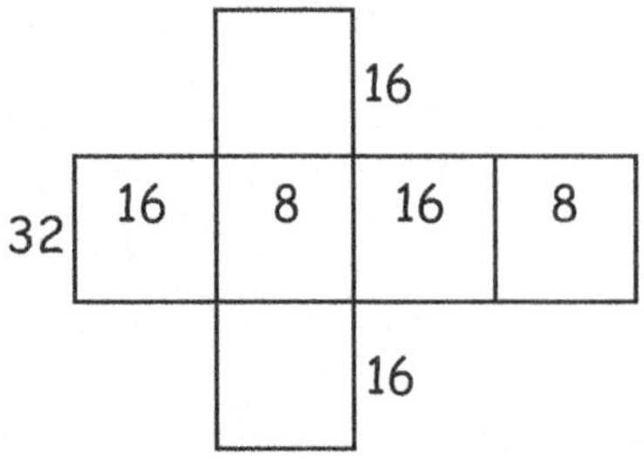

41. Find the missing ∠b for the below quadrilateral EFGH

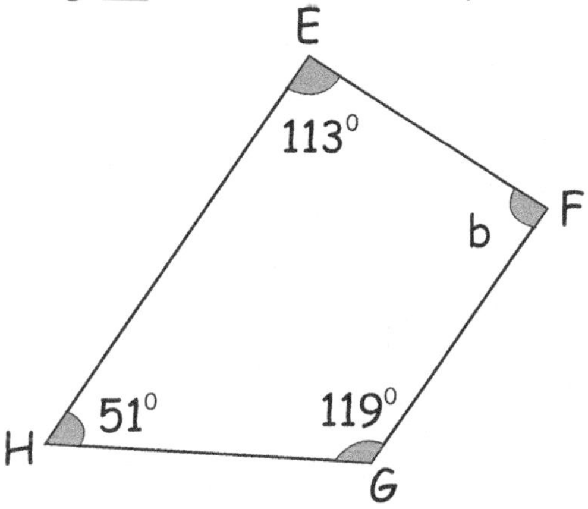

(A) 77°

(B) 90°

(C) 96°

(D) 86°

42. Find the area of the △ABC given below

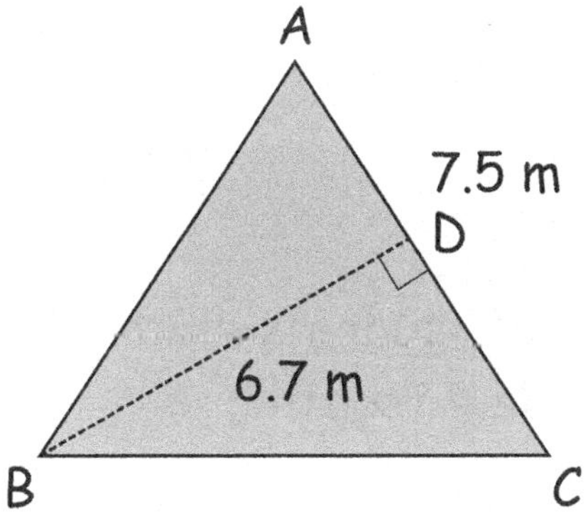

(A) 22.295 m^2

(B) 51.59 m^2

(C) 12.9 m^2

(D) 25.125 m^2

Tiny tots school has a total of 6,000 students in the year 2019 in various classes as represented in the below pi chart.

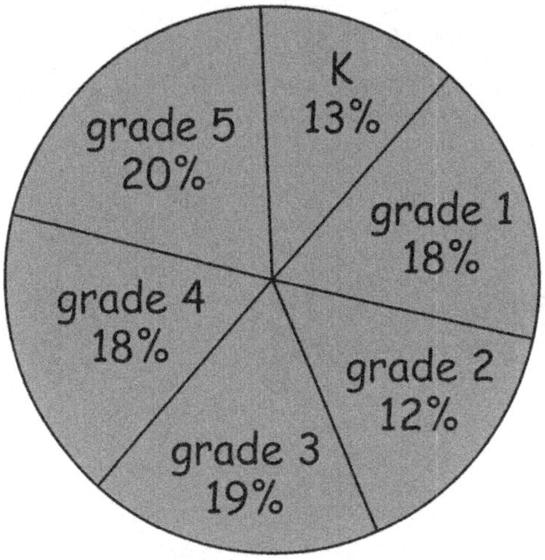

43. Find the difference between the total number of students studied in standard K and grade 3 ?

(A) 360 (B) 420

(C) 480 (D) 540

44. Find the average number of students studied in standard grade 1 , grade 2 and grade 4 together ?

(A) 1020 (B) 1280

(C) 630 (D) 980

45. How many yards are equal to 27 feet ?

(A) 3 (B) 9

(C) 12 (D) 8

46. To find which of the below we need highest to lowest value of the below given data set ?

(A) Mode (B) Mean

(C) Median (D) Range

47. Find the median of the below data set

4.82 , 1.29 , 4.3 , 2.02 , 1.26 , 2.07 , 1.84 , 2.07 , 1.43

(A) 2.02 (B) 2.86

(C) 1.84 (D) 2.07

48. Find the range of the below data set

42 , 43 , 44 , 44 , 50 , 50 , 51 , 55 , 59 , 60 , 64

(A) 9 (B) 5

(C) 22 (D) 11

49. Find the probability of getting a number less than 6 when a dice is rolled

(A) $\frac{1}{6}$

(B) $\frac{3}{4}$

(C) 1

(D) $\frac{5}{6}$

50. A city map has a scale of 1:9 cm. If building X and Building Y are 54 m apart then how far are they on the city map ?

(A) 48 cm

(B) 6 cm

(C) 8 cm

(D) 5 cm

Grade 6
SOL
Practice Test 1 - 3
Answer Keys

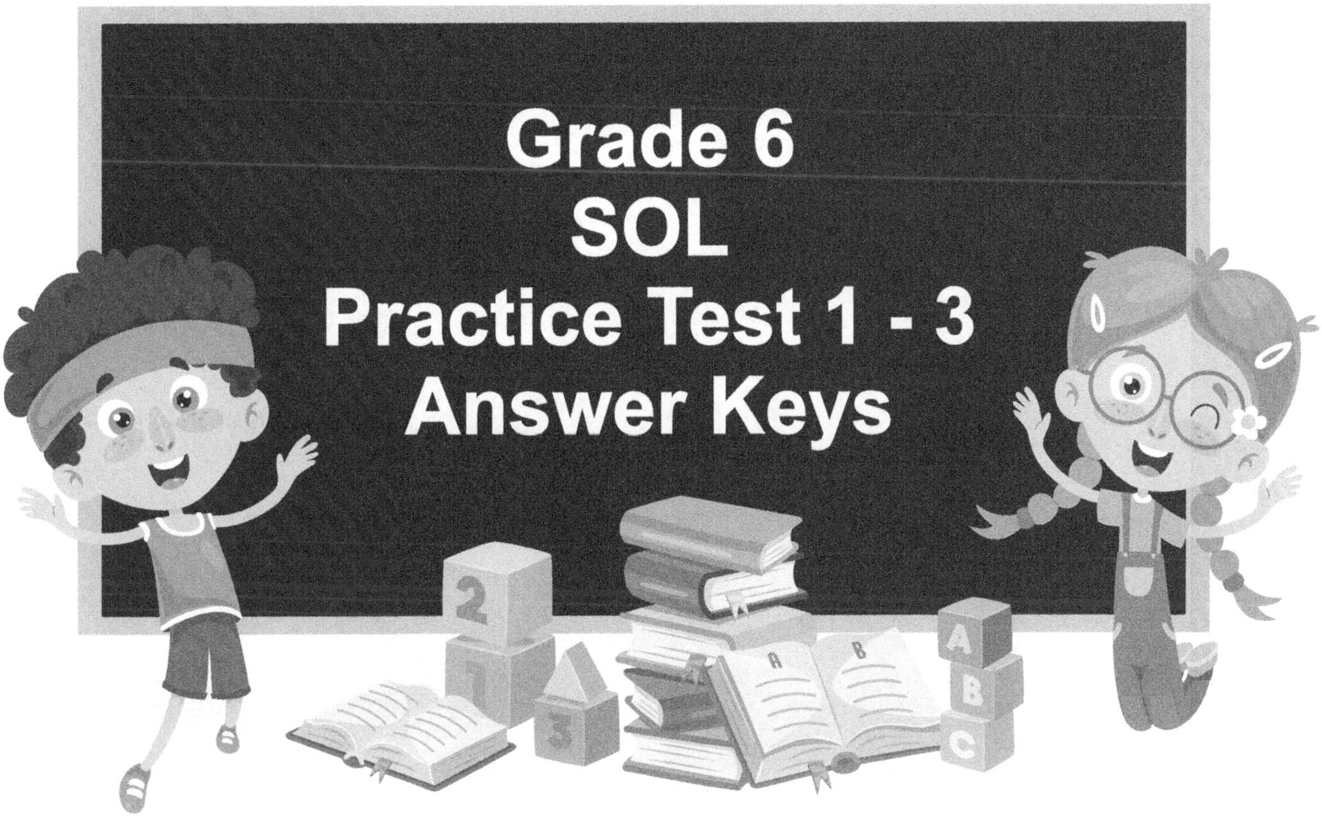

Answer Key Test - 1

1. C
2. C
3. D
4. A
5. B
6. D
7. A
8. C
9. D
10. D
11. C
12. A
13. D
14. B
15. D

16. A

17. C

18. B

19. A

20. D

21. A

22. C

23. D

24. B

25. C

26. A

27. B

28. B

29. D

30. A

31. C

32. C

33. D

34. C

35. B

36. A

37. B

38. D

39. D

40. A

41. C

42. B

43. A

44. C

45. C

46. A

47. C

48. D

49. A

50. (I) D

(II) C

Answer Key Test - 2

1. C
2. A
3. A
4. D
5. B
6. A
7. C
8. D
9. A
10. D
11. B
12. D
13. C
14. D
15. B

16. A

17. D

18. A

19. C

20. B

21. D

22. C

23. A

24. D

25. B

26. B

27. D

28. A

29. D

30. A

31. C

32. B

33. D

34. C

35. A

36. A

37. B

38. D

39. C

40. A

41. C

42. B

43. B

44. D

45. A

46. C

47. A

48. D

49. C

50. A

Answer Key Test - 3

1. A
2. C
3. B
4. D
5. C
6. B
7. C
8. D
9. B
10. C
11. A
12. B
13. B
14. A
15. C

16. B

17. D

18. C

19. A

20. A

21. B

22. D

23. C

24. C

25. D

26. A

27. D

28. A

29. C

30. C

31. A

32. D

33. B

34. A

35. A

36. D

37. D

38. C

39. C

40. C

41. A

42. D

43. A

44. D

45. B

46. D

47. A

48. C

49. D

50. B

GRADE 6
SOL
Score Calculation

Score calculation

If you get this many times correct :	Then your converted scale scor is :
0	000
1	217
2	249
3	268
4	282
5	293
6	303
7	311
8	319
9	325
10	331
11	337
12	343
13	348
14	353
15	357
16	362
17	366
18	371
19	375

20	379
21	383
22	387
23	391
24	395
25	399
26	403
27	407
28	411
29	415
30	419
31	423
32	427
33	432
34	436
35	441
36	445
37	450
38	456
39	461

40	467
41	473
42	480
43	487
44	496
45	505
46	516
47	531
48	550
49	582
50	600

Made in the USA
Monee, IL
05 November 2026

67914378R20052